Bruno Glaser · William I. Woods (Eds.)

Amazonian Dark Earths: Explorations in Space and Time

Springer

Berlin
Heidelberg
New York
Hong Kong
London
Milan
Paris
Tokyo

Bruno Glaser · William I. Woods (Eds.)

Amazonian Dark Earths:
Explorations
in Space and Time

With 46 Figures and 36 Tables

 Springer

Dr. Bruno Glaser
Institute of Soil Science and Soil Geography
University of Bayreuth
95440 Bayreuth
Germany

Professor William I. Woods
Department of Geography
Southern Illinois University
Edwardsville, IL 62026
USA

ISBN 3-540-00754-7 Springer-Verlag Berlin Heidelberg New York

Library of Congress Cataloging-in-Publication Data

Amazonian dark earths : explorations in space and time / B. Glaser, W. I. Woods (eds.).
p. cm.
Includes bibliographical references and index.
ISBN 3-540-00754-7 (alk. paper)
1. Anthropogenic soils–Amazon River Watershed. I. Glaser, B. (Bruno), 1966- II.
Woods, William I.

S599.3.A44A42 2004
631.4'9811–dc22

Springer-Verlag is a part of Springer Science + business Media
springeronline.com

© Springer-Verlag Berlin Heidelberg 2004
Printed in Germany

Cover Design: Design & Production, Heidelberg
Typesetting: Mitterweger & Partner GmbH, Plankstadt
31/3150WI – 5 4 3 2 1 0 – Printed on acid-free paper

Dedication

To WIM SOMBROEK:
friend, scholar, and the godfather
of Amazonian Dark Earths.

Preface

The idea for the volume first came about through a conversation the editors had at the Sustainable Management of Soil Organic Matter Conference in Edinburgh in September 1999. It developed with two symposia on Amazonian dark earths that were held in 2001 in conjunction with the Conference of Latin Americanist Geographers in Benicassim, Spain, and the Congress of the Brazilian Archaeological Society in Rio de Janeiro, Brazil, respectively, and culminated at the First International Workshop on Terra Preta Soils held in Manaus and Santarém, Brazil, in July 2002. As a comprehensive treatment of these distinctive anthropogenic soils has never been published, we decided to select papers from these symposia and develop an edited volume. The result contains the efforts of an international group of distinguished scholars from the disciplines of anthropology, archaeology, biology, geography, and soil science. The 15 chapters of this volume provide an array of interesting and complementary interpretative stances developed from a diverse body of investigative methodologies. The reader will note that there are some inconsistencies in terminology and differences in interpretation among the chapter presentations. However, the editors purposely allowed these to remain and retained as much as possible of the authors' own words, since we felt that it was important to maintain the flavor of the symposium atmosphere in this volume and consequently did not intentionally force standardization upon the authors.

Interest among scholars from a variety of disciplines on the topic of the dark earths has recently been enhanced by articles in many journals, including *Science* and *Naturwissenschaften*. The public has also been informed through a noteworthy exhibition at the British Museum entitled "The Unknown Amazon" in which the significance of the dark earths was emphasized and recent articles in such popular magazines as *Der Spiegel* and *The Atlantic Monthly* (both in March 2002 issues) and in many newspaper pieces and online notices and commentary. During the 2002 Manaus workshop and field conference, a documentary with special emphasis on Amazonian dark earths was filmed and subsequently broadcast that December on BBC 2. Additional programs have appeared on German television more recently. The issues posed by the dark earth soils from the past potentially provide a rich resource for the future of Amazonia and, indeed, the world. It is our hope that this volume will help the reader attain a better understanding of these soils and the efforts toward resolving the basic questions of their past, present, and future.

The editors thank all the participants of the three symposia for their invaluable contributions toward Amazonian dark earth research. These individuals provided both individually and collectively the true spirit of shared scholarship that makes these investigations such a joy to be a part of.

We want to thank Dr. Dieter Czeschlik, the Editorial Director for the Life Sciences at Springer-Verlag, and his assistant, Mrs. Anette Lindquist, who steered us through the publication process with plenty of patience and expert advice. A special acknowledgement is given to Andrew Martignoni, Jr. of the SIUE Office of Contract Archaeology who prepared the final formatting of this volume's figures. We are also indebted to the German Research Foundation as well as the Southern Illinois University Edwardsville Graduate School and College of Arts and Sciences for their support of this work and volume.

Finally, we wish to thank our colleagues, friends, and families for their patience and support of our work, without which we would not have been able to present this publication.

Bayreuth and Edwardsville,   Bruno Glaser and William I. Woods
March 2004

Contents

Towards an Understanding of Amazonian Dark Earths

WILLIAM I. WOODS[1] and BRUNO GLASER[2]

1.1
Anthrosols and Amazonia

By way of introduction to the topic and to provide a basis for the context of this volume, we provide the following short discussion. During the last century, a number of researchers looked seriously at development scenarios for various parts of the world and attributed great importance to natural soils and their management or mismanagement. In the case of Amazonia, the extremely poor soils present were cited as the fundamental cause for lack of cultural attainment. In general, soils were used deterministically as an explanatory mechanism when they presumably suited a preconceived notion of development or decline, but little in-depth study was undertaken regarding which soils were actually present on the microscale and how these had been affected by different management strategies over time.

Soils, along with energy from the sun, water, and the cultivated plants themselves, constitute the major environmental resources for agricultural societies and are, therefore, critical to discussions of sustainability. In the Amazon region, often depicted as a "Counterfeit Paradise" or "Green Hell," the highly weathered, very acidic soils of the *terra firme* (upland settings) are thought of as extremely forbidding. With few available nutrients and having extremely high aluminum concentrations, one could not imagine a worse regime for productive agriculture, particularly when associated with nucleation of population. Indeed, even in the *várzea* (floodplains) with somewhat better soils, crop production has been seen as a risky endeavor because of the unpredictability of the flood regime.

However, it is a matter of scale when dealing with humans and the environment; and for soils, this is particularly true. Continental or regional depictions simply are not appropriate. Rather, one needs to look at the microscale and here one finds great variety in and enormous pre-Columbian modifications to the soil landscape and unexpected answers to questions of sustainability. Most Amazonian soils have been modified and manipulated, intentionally and unintentionally, directly and indirectly, negatively and posi-

[1] Department of Geography, Southern Illinois University Edwardsville, Edwardsville, Illinois 62026, USA
[2] Institute of Soil Science and Soil Geography, University of Bayreuth, 95440 Bayreuth, Germany

tively by the combined activities of the region's aboriginal inhabitants over time.

When dealing with humans and their effects on the land, enormous complications come into play due to the extreme heterogeneity of the types, amounts, and distribution of inputs and withdrawals over time. Anthropogenic soils exhibit these problems in excess at all scales of inquiry and tend to form a continuum of expression within any given micro-habitat. The reality of our aboriginal farmers, their basic decisions, and the majority of outcomes were at the level of the individual and his or her family and household. The point of articulation with the environment at this scale was narrow, mainly including only the limited zone of exploitation surrounding the place of habitation. The result was an extremely heterogeneous mix of adaptive modifications whose resulting soils vary in the extreme.

One tends to associate land degradation with tropical soil use, particularly in Amazonia, where numerous instances from localized deflation and nutri-

Fig. 1.1a. Typical profiles of Latosol

ent depletion have been documented. However, questions have been raised about this overly dismal viewpoint and it has been proposed that complex societies with large, sedentary populations were present for at least a millennium before European contact. Early explorers described dense villages extending for kilometers along river bluff edge settings with roadways linking these to settlements in the interior. Often associated with the presumed village locations are soils termed "Indian black earth," or *terra preta* (Fig. 1.1). The heightened fertility status of these soils, generically termed "dark earths" herein, has long been recognized by the indigenous inhabitants of the region, as well as by current colonists. Throughout Amazonia, dark earths occur in a variety of landscape contexts, in circumscribed patches ranging in size from less than 1 ha to many square kilometers. The most nutrient-demanding crops are often planted on dark earth sites. When not cleared for agriculture, a distinctive vegetation structure and species composition are recognized, and the unique and abundant assemblage of useful wild and semi-domesticated plant species occurring on them is exploited.

Fig. 1.1b. Typical profiles of *terra preta* soils

Currently, most researchers believe that these soils formed in cultural deposits created through the accretion of waste and occupation debris around habitation areas. The great enrichment of essential plant nutrients such as phosphorus, calcium, and potassium found in many of these soils certainly supports such a conclusion. Theories about the anthropogenic origins of the dark earths also include a burning scenario, whether from fires within the habitation area or from those used to clear forest for fields and clear fields of choking weedy growth. Fire contributes charcoal and ash, which increase soil pH, thereby suppressing aluminum activity toxic to plant roots and soil microbiota. The consequent increase in microbiological activity adds colloidally sized organic decomposition products to the soil matrix. These, along with the by-products of incomplete combustion, provide charged surfaces largely absent in the local soils and increase nutrient retention capacity, setting up a synergistic cycle of continued fertility.

1.2
Volume Overview

In the lowland humid portion of the Amazon Basin, intensively weathered nutrient-poor soils such as Oxisols, Ultisols, and Acrisols predominate. Embedded in this landscape of infertile soils are patches of dark earths with huge carbon- and nutrient-rich A horizons providing sustainable land use. The following 14 chapters of this book investigate various aspects of the dark earth phenomenon.

In Chapter 2, Glaser, Zech, and Woods trace the development of scientific knowledge of the dark earths from the first descriptions by observers in the 1870s. Although these early accounts supposed an anthropogenic origin, later discussions presented a variety of geogenic-origin hypotheses. More intensive investigations starting in the 1960s have provided a clearer picture of the widespread distribution of these soils. Analyses of their chemical and physical parameters have shown that the dark earths were formed in situ by the activities of pre-Columbian Indians. Recent work at the molecular level has revealed that the dark earths contain considerable amounts of charring residues which are known to persist in the environment for centuries. However, the specific sources for the nutrient enrichment found in many of these soils are still unclear. These authors conclude that future investigations should focus on the identification of land-use practices of the pre-Columbian population and on the implementation of this knowledge in order to reproduce sites of enriched soils.

Kern, Costa, and Frazão in Chapter 3 also review the scientific background for Amazonian dark earth studies and similarly come to the conclusion that these soils are the product of aboriginal settlement activities. They provide geochemical data that identify chemical associations of elements that were added to the soil by past human occupations and represent the geochemical signature for modified regional soils. These data also allow the hypothetical determination of the settlement pattern in a prehistoric site with *terra preta*, as well as its evolution over the centuries. Ongoing research concerns the chemical composition of plants like palm trees and manioc, mostly to detect the origin of the high values of Zn and Mn which occur in the *terra preta*. It is already known that ceramics present in archaeological black earth (ABE) contain very high contents of some chemical elements.

German in Chapter 4 outlines a method for the geographical characterization of Amazonian dark earths. The method is employed to map and characterize sites on the Rio Negro, where the environment has been portrayed as a constraint to human subsistence and sedentary occupation. The degree of anthropogenic soil modification is estimated from total phosphorus content, epipedon depth, and soil color. Findings show that human settlements over time were concentrated along major waterways, at the headwaters of smaller tributaries, and at the confluence of two rivers. Rather than corresponding to distributions of a single "limiting" resource, settlement areas are shown to provide a number of advantages to human inhabitants, including access to multiple resources, political-economic control and defense.

In contrast to the Brazilian Amazon, few *terra preta* sites have yet been identified in the Upper Amazon. In Chapter 5, Coomes describes the characteristics and human uses of a site on a relic meander island complex in the Amazon River lowlands of northeastern Peru. Local residents refer to the site as the *'yarinales'* (after the presence of the ivory nut or *yarina* palm). The soils of the *yarinales* are distinct from those of the Amazon River floodplain and the Ultisols of the *terra firme*, by elevated phosphorus concentrations, dark color, and high sand content. Pot sherds and stone implements suggest prehistoric occupation of the site. Since the late nineteenth century, farmers have practiced one of the most productive and profitable forms of lowland swidden-fallow agroforestry yet described for the Amazon Basin. Crop–fallow cycles are tailored to the elevation of the land relative to periodic floodwaters. Residents also rely on the *yarinales* for the extraction of non-timber forest products as well as medicinal plants and for hunting. The common occurrence of relic riverine features in the Peruvian Amazon suggests that similar sites of high agricultural potential are to be found in the Upper Amazon.

Prehistoric migrations out of the central Amazon may have been motivated by a population explosion made possible by dark earth technology. In Chapter 6, Myers analyzes sites from the Upper Amazon to determine (1) if dark earth technology accompanied immigrants from the central Amazon, and (2) if there was an indigenous development of dark earth technology. His analysis reveals that immigrants were not frequently associated with dark earth sites and that dark earth sites are sometimes associated with indigenous peoples. Thus it seems unlikely that black earth technology was the key innovation that permitted population growth and the subsequent diaspora from the central Amazon to its principal tributaries.

Ruivo, Cunha, and Kern studied the mineral, elemental chemical, and organic components of soils from five sites located in the Ferreira Penna Scientific Station (ECFPn), State of Pará, Brazil, and report on this work in Chapter 7. Four sites contained Yellow Latosols, while the fifth was composed of an ABE. They determined that chemical differences between the sites' soils were due not only to patterns of past human enrichment, but also to variations in soil texture and drainage.

Lehmann et al. in Chapter 8 tested the applicability and suitability of a sequential extraction technique for phosphorus for characterizing anthropogenic dark earths. They determined the P distributions in different soil pools for Anthrosols and for Ferralsols from central Amazonia with high, low, and no P fertilization and demonstrated that the dark earths contained 2 to 30 times more total P than unfertilized Ferralsols, less recalcitrant P than both the fertilized and unfertilized Ferralsols, and less readily extractable and plant-available P than the highly fertilized Ferralsol. Furthermore, the prevalence of calcium-bonded phosphate gives reason to hold fish waste responsible for the high P contents of the dark earths, because bones contain high amounts of P in a similar form.

Neves et al. in Chapter 9 present the preliminary chronological data on *terra preta* formation in three archaeological sites around the area of confluence of the Negro and Solimões rivers. Like so many terrace or bluff settings adjacent to floodplains in the Amazon, one finds in the area several large archaeological sites with dense ceramic deposits associated with deep *terra preta*. Stratigraphic excavation of these sites produced 44 charcoal samples that were radiocarbon dated. The results of the dates indicate that, contrary to our preliminary expectations, *terra preta* formation was in these cases a fairly rapid process. The authors offer an interpretation of this interesting and quite unexpected finding.

Anthropogenic dark earths are widespread in the uplands of Amazonia, in patches covering 1 ha or less up to several hundred hectares. The blacker form (*terra preta*) seems to have developed from pre-European village middens. The lighter, dark brown form (*terra mulata*), which is much more extensive, is believed to be the product of intensive cultivation practices. In Chapter 10, Denevan concentrates on the *terra mulata* and suggests that its high content of black carbon particles is the residues of frequent, incomplete burning by pre-European people. He describes such practices for present Indian groups in Amazonia and suggests that, once established, these superior dark earth soils were particularly attractive to farmers because nutrient-demanding crops such as maize could be grown successfully on them. Thus, these soils acted as further stimulus for both maintaining settlement and cultivation in these sites and returning to them when/if there were periods of abandonment, continuing to the present.

The anthropogenic origin of *terra preta*, a strongly weathered soil with a thick humic A horizon, high soil fertility, and potsherds, is now well accepted. However, it remains unknown how the present advantageous soil properties are related to the pre-Columbian anthropogenic activities that also created the thick humic A horizon. Therefore, in the study presented in

Chapter 11, Glaser, Guggenberger, and Zech examine physical and chemical properties of *terra preta* soils in comparison to the generally poor surrounding soils and relate them to site-specific factors. They conclude that the *terra preta* phenomenon is a product of both anthropogenic and pedogenic factors, in particular the clay content. The human impact decreased with increasing profile depth, indicating an influence predominating on the soil surface.

In Chapter 12, Silva and Reballato present a progress report on research that aims to understand the existing relationship between the dynamics of space use in an ethnographic context and the cultural formation processes leading to *terra preta*. They describe the way in which the indigenous population Asurini do Xingu utilizes the village space in their daily activities. Within the community their research has identified the different configurations of the activity areas (public, domestic, and disposal areas) and, at the same time, called attention to the fact that the disposal areas are the most favorable to the formation of *terra preta*.

Recent field studies indicate that loci of *terra preta* are often surrounded by bands of soils with dark brown rather than black color and exhibit a near absence of cultural artifacts, but still maintain the same high amount of soil organic matter (SOM). They are locally called *terra mulata*, and would appear to be the lasting result of intensive agricultural use by erstwhile Indian communities. The SOM of both the *terra preta* and the *terra mulata* is very stable and at the same time more re-active than the SOM of the original soils under primary forest cover. In Chapter 13, Madari, Sombroek, and Woods describe the development of an interdisciplinary and inter-institutional project designed to establish the reasons for the sustainability of these soils; to obtain understanding on the land-use practices of the former Indian tribes that led to such a lasting enrichment, and to develop a replicating land management package for the benefit of more sustainable and more productive food production for smallholder farmers.

Steiner, Teixeira, and Zech in Chapter 14 describe a newly discovered agricultural practice in which the residues of charcoal production are used to improve soil quality in the vicinity of Manaus and in other parts of Brazil. The soil charcoal amendments maintain soil fertility and transfer carbon from the atmosphere into stable SOM pools, which can improve and maintain the productivity of highly weathered infertile tropical soils. The agricultural practice of slash and char produces charcoal from the aboveground biomass instead of converting it to carbon dioxide through burning. The authors come to the conclusion that slash and char practiced as an alternative to slash and burn throughout the tropics could be an important step towards sustainability in tropical agriculture.

Steiner et al. in Chapter 15 report that frequent findings of charcoal and highly aromatic humic substances suggest that residues of incomplete combustion of organic material (black carbon, pyrogenic carbon, charcoal) are a key factor for the persistence of SOM and maintenance of fertility in *terra preta* soils. However, charred organic matter and black carbon in terrestrial soils have not been evaluated regarding their importance for nutrient supply and retention in soil. Microorganisms transform and cycle organic matter and plant nutrients in the soil and are sinks (during immobilization) and sources (during mineralization) of labile nutrients. The microbial biomass can serve as an indicator of the effect of different management practices. In this chapter, a field experiment designed to measure the effects of organic matter amendments to soil (charcoal, chicken manure, compost, and litter) was tested in central Amazonia with respect to microbial respiration. The methodology, results, and preliminary interpretations are discussed in regard to measures of soil fertility and quality.

1.3
Importance of the Dark Earth Investigations

Many questions are still unanswered with respect to the origin, distribution, and properties of the Amazonian dark earths and further exploration of these soils is highly significant for a number of reasons, including:
1. The dark earths contain invaluable information about settlement and environmental management strategies and associated productivity and sustainability considerations during the pre-European period.
2. Present utilization of these soils provides an important resource for food production within Amazonia.
3. By recreating the high and sustainable soil fertility conditions of the past, smallholder farmers should be able to improve their standard of living through more productive and environmentally friendly crop production.
4. The long-term persistence of the elevated SOM in these soils has important implications for carbon sequestration, a major controlling factor for global climate change.

2 History, Current Knowledge and Future Perspectives of Geoecological Research Concerning the Origin of Amazonian Anthropogenic Dark Earths (*Terra Preta*)

BRUNO GLASER[1], WOLFGANG ZECH[1], and WILLIAM I. WOODS[2]

2.1
Introduction

2.1.1
The Paradox of the Existence of *Terra Preta*

Soil degradation is one of the most severe problems of land use in the lowland humid tropics (Zech 1997) largely due to the fact that soil organic matter (SOM) is mineralized rapidly under the optimum growth temperatures for micro-organisms (Tiessen et al. 1994). However, SOM is of particular importance for sustainable agricultural use of the heavily weathered tropical soils. It contributes substantially to nutrient supply, cation exchange capacity (CEC), and to a favorable soil structure (Ross 1993). A loss of SOM after slash-and-burn and other agricultural and pastoral land uses progresses the soil degradation of many tropical soils, resulting in infertile soils after a few years of cultivation. Soil amelioration by application of mineral fertilizers or compost is often unaffordable for poor smallholder farmers or remains ineffective due to the low CEC of the soils or the lack of knowledge about the nutrient release from organic fertilizers (Tiessen et al. 1994).

In the Brazilian Amazon Basin, too, intensively weathered, nutrient-poor soils such as Oxisols, Ultisols, and Acrisols dominate. However, embedded within this landscape of infertile soils are patches of black earths with huge carbon and nutrient-rich A horizons providing a sustainable land use. These soils are locally known as Indian Black Earths or *terra preta* (*do índio*). The occurrence of ceramics and charcoal indicates that these soils were anthropogenically influenced. However, the question whether *terra preta* soils have been formed by human activities or if sustainably fertile soils already existed prior to the pre-European habitation of dark earth expanses is as old as the investigation of *terra preta* itself. Until recently, the origin of *terra preta* has been controversially discussed.

[1] Institute of Soil Science and Soil Geography, University of Bayreuth, 95440 Bayreuth, Germany
[2] Department of Geography, Southern Illinois University Edwardsville, Edwardsville, Illinois 62026, USA

2.1.2
Geogenic Versus Anthropogenic Formation of *Terra Preta*

Herbert Smith (1879) was one of the earliest to describe *terra preta* or "black land." On areas in the vicinity of the Tapajós River and the town of Santarém in the Lower Amazon he reported that fragments of Indian pottery were so abundant in some places on the dark earth sites that they almost covered the ground. Fredrick Katzer (1903) working in the same area three decades later came to similar conclusions about the correlation between pre-European habitation activities and the dark earths. However, Barbosa de Farias (1944) proposed that *terra preta* sites were already fertile before they were settled by the native population. Thereafter, a sequence of geogenic origins was proposed. For instance, volcanic (Hilbert 1968) and fluvial (Zimmermann 1958; Franco 1962) sedimentation were suggested. In response, Ranzani et al. (1962) suggested plaggen deposition, thus coming back to the human factor in the genesis of these soils. Since then, an anthropogenic origin for the dark earths has been favored (e.g., Sombroek 1966; Hilbert 1968; Zech et al. 1979; Smith 1980; Zech et al. 1990; Sombroek et al. 1993; Glaser 1999; Woods and McCann 1999; Glaser et al. 2000; Woods et al. 2000; Glaser et al. 2001).

In summary, four different theories of *terra preta* formation have been postulated: (1) volcanic sedimentation; (2) fluvial deposition; (3) anthropogenic plaggen deposition; and, (4) anthropogenic in situ formation. With the current pedological knowledge that *terra preta* soils and the surrounding infertile soils have a similar mineralogical composition, all geogenic hypotheses of *terra preta* formation can be rejected. For instance, upon volcanic sedimentation one would expect the deposition of tephra layers coupled with a different particle-size distribution, the occurrence of volcanic glasses and their weathering products (allophanes), and different heavy minerals (Gillespie et al. 1992; Zech et al. 1996) in *terra preta* soils. A fluvial deposition would also result in a different texture, leading to an enrichment of the finer silt and clay fractions in *terra preta* soils when compared with their background counterparts. However, this has never been reported (Zech et al. 1979; Sombroek et al. 1993; Glaser et al. 2004). Instead, *terra preta* soils have sometimes been found to contain slightly higher amounts of sand in the upper horizons, indicative of an anthropogenic influence (Smith 1980; Glaser et al. 2004). A plaggen application changes the soil texture and leads to raised soil surfaces (Conry 1974) and both have not been reported for *terra preta* soils (Zech et al. 1979; Sombroek et al. 1993; Glaser et al. 2004). Due to these facts, soil scientists favor the in situ genesis hypothesis of *terra preta* formation, although some dark earth sites were heavily disturbed by mound formation (Petersen et al. 2001).

2.2
History of Geoecological *Terra Preta* Research

During the 1960s and 1970s, *terra preta* sites throughout the Amazon Basin were mapped and investigated with respect to soil physical and chemical parameters supporting the anthropogenic origin of *terra preta* soils. It was shown that these soils contain higher amounts of organic carbon, nitrogen, phosphorus, and other important nutrients for plant growth compared to surrounding soils (Sombroek 1966; Hilbert 1968; Zech et al. 1979). In addition, due to neutral or weak acid pH conditions, these nutrients are more available to plants compared to those in the strongly acid parent soils. In the 1980s *terra preta* had become widely viewed as a kind of kitchen-midden, which has acquired its specific fertility from urine and excrement, household garbage, and the refuse of hunting and fishing (Smith 1980; Zech et al. 1990). Due to these facts, *terra preta* soils are favorable for high-yielding agricultural production under humid tropical conditions.

Zech et al. (1990) investigated the quality of the organic matter of *terra preta* soils in comparison with surrounding infertile soils using ^{13}C nuclear magnetic resonance spectroscopy (^{13}C NMR). The most striking difference was the higher contribution of aromatic and carboxyl carbon of *terra preta* soils, leading the authors to the conclusion that the development of stable aromatic SOM was decisive for the sustainable fertility of *terra preta* soils. This hypothesis was the basis for the work of Glaser (1999) who explored intensively the properties and stability of the SOM of *terra preta* soils.

The work of Glaser (1999) was based on the three theoretical mechanisms of SOM stabilization in mineral soils (Christensen 1992): (1) chemical resistance due to polymerization or selective enrichment of stable compounds; (2) organo-mineral stabilization due to binding of organic matter to oxides or clay minerals; and (3) physical stabilization of easily decomposable organic matter or microorganisms due to entrapment into soil aggregates. Owing to the fact that all these processes could act simultaneously or successively in soil, methods had to be applied which could discriminate between the different stabilization mechanisms.

One of the most important differentiation criterions between the different stabilization mechanisms is the scale level. Chemical resistance occurs at the molecular level and can be traced by the analysis of the chemical structure of different carbon species (Kögel 1987). The stability of *terra preta* SOM could be caused by selective enrichment of stable, aromatic compounds during "humification" (Zech et al. 1990; Kögel-Knabner 1993). Also, an enrichment of charring residues (pyrogenic carbon) as the source of stable, aromatic structures was speculated (Haumaier and Zech 1995). To investigate the organo-mineral stabilization theory, the SOM distribution and composition of organo-mineral primary particles with different particle size and organo-mineral secondary particles with different density were looked at. To characterize physical stabilization of SOM, soil aggregates were studied using UV-photo-oxidation techniques similar to those used by Skjemstad et al. (1993, 1996).

2.3
Current Geoecological Knowledge About *Terra Preta*

2.3.1
Nutrient and Organic Matter Levels in *Terra Preta* Soils

Enhanced fertility due to high nutrient levels such as N, P, and Ca is characteristic of *terra preta* soils (Sombroek 1966; Zech et al. 1990; Glaser 1999;
Smith 1999; Glaser et al. 2000, 2001, 2004). These soils are very popular
among local farmers and are preferably used to produce cash crops such as
fruits and vegetables, which give higher yields and more rapid plant development than on surrounding infertile soils (Woods and McCann 1999; German
2001). Fallow periods on Oxisols/Ferralsols usually last 8–10 years, whereas
fallows on *terra preta* soils effectively restoring their fertility can be as short
as 6 months (German 2001). The cropping period on *terra preta*, however, is
generally shorter than on adjacent Oxisols (4 months to 2 years in comparison to 2–3 years), respectively, due to weed invasion on the fertile soils (German 2001).

 Terra preta soils contain not only higher levels of available nutrients, but
also higher amounts of SOM. The total organic carbon stocks can be as high
as $250\,Mg\,ha^{-1}$ in the agronomically important soil depth of 0–0.3 m (Glaser
et al. 2001) and $500\,Mg\,ha^{-1}$ up to 1 m soil depth (Glaser 1999). In comparison, adjacent Oxisols/Ferralsols may contain only 100 and $149\,Mg\,C\,ha^{-1}$ in
0–0.3 m and up to 1 m soil depth, respectively. Therefore, carbon sequestration is about three to four times higher in the *terra preta* soils compared to
adjacent soils. However, the total carbon sequestration on a landscape level is
unclear and should be subject to further research.

2.3.2
Stability of the Organic Matter in *Terra Preta* Soils

Recent investigations (Glaser 1999; Glaser et al. 2000, 2001) have shown that
charred residues from incomplete combustion of organic material (i.e., black
carbon, pyrogenic carbon, charcoal) are responsible for maintaining high
levels of SOM and available nutrients in *terra preta* soils. In these soils large
amounts of pyrogenic carbon indicate a high and prolonged input of carbonized organic matter probably due to production of charcoal in hearths,
whereas only low amounts of charcoal are added to soils as a result of forest
fires and slash-and-burn techniques (Fearnside et al. 1999; Fearnside 2000).
Up to 70 times more pyrogenic carbon (charcoal) is contained in *terra preta*
than the surrounding soils (Glaser 1999; Glaser et al. 2001). It is assumed that
pyrogenic carbon persists in this environment over at least centuries due to
its chemical stability caused by the aromatic structure. The complex chemical
structure makes the compound also resistant to microbial degradation (Seiler

and Crutzen 1980; Goldberg 1985; Schmidt et al. 1999). This assumption was emphasized by ^{14}C ages of 1,000–2,000 years of this carbon type (Glaser et al. 2001). Oxidation during this time produced carboxylic groups on the edges of the aromatic backbone, which increased the nutrient retention capacity (Glaser et al. 2000). It was concluded that pyrogenic carbon found in these anthropogenic soils not only acts as a significant carbon sink, but also is a key factor in maintaining the sustainable fertility of *terra preta* soils.

Investigations concerning the physical and organo-mineral stabilization of SOM showed that SOM is mainly stabilized via sorption to mineral surfaces, whereas physical stabilization via entrapment into the interior of aggregates accounts only for about 20 % in *terra preta* soils compared to 10 % in adjacent soils (Glaser 1999). Therefore, besides the occurrence of recalcitrant SOM in pyrogenic forms, the stability of *terra preta* SOM can be partly explained by physical stabilization in aggregates. Additionally, higher SOM amounts in *terra preta* soils favor soil aggregation (Glaser 1999).

2.4
Origin of *Terra Preta* Soils and Future Research Perspectives

2.4.1
Origin of Enhanced Organic Matter Levels

Whereas there is no doubt about the accumulation of pyrogenic carbon as a source of stable SOM in *terra preta* soils (Glaser et al. 2001), its specific origin is still a matter of speculation. Although it was shown that principally it is possible to accumulate the amounts of pyrogenic carbon found in *terra preta* soils by about 25 natural fires occurring at the same site (Glaser et al. 2001), this scenario does not seem very probable. As it is assumed that natural forest fires have occurred several times throughout the whole area of the Amazon Basin, then all of Amazonia should be a *terra preta* by this hypothesis. Another calculation based on slash-and-burn agriculture gave evidence that only a single burn could accumulate pyrogenic carbon amounts found in *terra preta* soils (Glaser et al. 2001). Even this scenario seems improbable as *terra preta* soils should still be formed today under shifting cultivation, a fact that has never been reported.

More likely than natural and slash-and-burn fires as a source of pyrogenic carbon in *terra preta* soils are the low-heat smoldering domestic fires commonly used by the native population for cooking and heating (Smith 1980, 1999). To simulate such smoldering, pine wood was isothermally charred in a muffle furnace at 300 °C, yielding about 20 % of charcoal (Glaser et al. 1998). Supplying pyrogenic carbon from this source to the soil, about 600 Mg pine wood has to be charred, producing more than the measured black carbon stocks of *terra preta* soils per hectare. It is difficult, however, to calculate the 'true' charring frequency or amount of wood or organic debris necessary to produce a specific quantity of pyrogenic carbon. This is due to the different properties and pyrogenic carbon yields of the materials used and to the continuum of pyrogenic carbon which is produced under different charring conditions (Schmidt and Noack 2000).

WinklerPrins (2002) found that by far the dominant form of soil management in current Amazonian urban gardens was the creation and use of *terra quiemada* (burned earth). Garden managers sweep their gardens daily, an activity that was so common that most did not mention this as 'management' at all. Sweeping the garden of fallen debris is in the same category as sweeping your house. This garden debris is collected in one particular place within the house-lot. There it would be burned approximately once a week. The ashes and combusted remains of the leaves and twigs gradually mixed in and darkened the sandy soil of the garden. This mixture is then placed around favored trees or in pots for growing vegetables to act as a fertilizer. Periodically, after several years, the burning site is moved to another part of the house lot and a specific tree or shrub planted on the old site. With time, carbonized residues accumulated in this way, coupled with those from hearths, processing facilities, and other activities involving burning would produce appreciable amounts of pyrogenic carbon.

2.4.2
Origin of Enhanced Nutrient Levels

Up to now, only speculations exist about the origin of the high nutrient levels of *terra preta* soils and the interaction with the high SOM levels. Glaser (1999) investigated ecologically relevant N and P pools by means of physical, chemical, and biological methods and found that N is stabilized especially in silt and clay fractions whereas P was higher in sandy *terra preta* soils, suggesting different sources for N and P. It is assumed that the high P stocks are derived from excrements and bone residues (Zech et al. 1990; Glaser 1999; Smith 1999).

Glaser (1999) identified 30% of total N in *terra preta* soils to which 18–25% contributed amino acid N, 4–7% amino sugar N, and 1–2% inorganic N. These values were similar to the relative composition of N sources in surrounding non-*terra preta* sites, suggesting that the principal sources of nutrient input into the soil were equal for *terra preta* and adjacent soils. Thus, the only difference seems to be the quantity of nutrient input. It was speculated that the major part of the high amounts of unknown N in *terra preta* soils consisted of heterocyclic N and it is known that pyrogenic carbon contains such N forms.

In this context, soil fertility might be causally linked to the accumulation of pyrogenic carbon. There might be even a positive feedback loop: High primary productivity of the *terra preta* soils promotes the accumulation of pyrogenic carbon upon vegetation burning, thus functioning as an effective CO_2 sink in terrestrial environments. The high pyrogenic carbon contents in the soil, however, are slowly abiotically oxidized, which elevates cation exchange capacity and thus soil fertility. This again promotes primary plant productivity and so on. This theory is supported by the higher carbon and nitrogen mineralization potential of *terra preta* soils in comparison with surrounding infertile soils, promoting a continuous nutrient supply for plant growth (Glaser 1999) if the system is managed in a subsidiary instead of an exploitive manner.

2.5
Conclusions

The existence of *terra preta* soils underlined by the above-mentioned results proves that it is possible to convert degraded Oxisols into sustainable fertile soils. Future investigations should focus on the identification of land-use practices of the pre-Columbian population and on the implementation of this knowledge in order to produce new *terra preta* (*"Terra Preta Nova"*) sites.

References

Barbosa de Farias J (1944) A cerâmica da tribo Uaboí dos rios Trombetas e Jamundá – Contribuição para o estudo da arqueologia pré-histórica do Baixo Amazonas. In: C.N.d. Geografia (ed) 9th Congresso Brasileiro de Geografia, 1940, Anais III, Rio de Janeiro, pp 141–165

Christensen BT (1992) Physical fractionation of soil and organic matter in primary particle size and density separates. Adv Soil Sci 20:1–90

Conry MJ (1974) Plaggen soils, a review of man-made raised soils. Soils Fertil 37:319–326

Fearnside PM (2000) Global warming and tropical land-use change: greenhouse gas emissions from biomass burning, decomposition and soils in forest conversion, shifting cultivation and secondary vegetation. Clim Change 46:115–158

Fearnside PM, Graca PML, Nilho NL, Rodrigues FJA, Robinson JM (1999) Tropical forest burning in Brazilian Amazonia: measurement of biomass loading, burning efficiency and charcoal formation at Altamira, Pará. For Ecol Manage 123:65–79

Franco E (1962) As "Terras Pretas" do Planalto de Santarém. Rev Soc Agrôn Veterinários Para 8:17–21

German L (2001) The dynamics of terra preta: an integrated study of human–environmental interaction in a nutrient-poor Amazonian ecosystem. University of Georgia, Athens, Georgia, 336 pp

Gillespie R, Hammond AP, Goh KM, Tonkin PJ, Lowe DC, Sparks RJ, Wallace G (1992) AMS dating of a late quaternary tephra at Graham's Terrace, New Zealand. Radiocarbon 34:21–27

Glaser B (1999) Eigenschaften und Stabilität des Humuskörpers der Indianerschwarzerden Amazoniens. Bayreuther Bodenkundl Ber 68:196 pp

Glaser B, Guggenberger G, Zech W (2004) Past anthropogenic influence on the present soil properties of anthropogenic dark earths (terra preta) in Amazonia (Brazil). In: Glaser B, Woods WI (2004) Amazônian dark earths: Explorations in space and time. Springer, Berlin Heidelberg New York, 250 pp

Glaser B, Haumaier L, Guggenberger G, Zech W (1998) Black carbon in soils: the use of benzenecarboxylic acids as specific markers. Org Geochem 29:811–819

Glaser B, Guggenberger G, Haumaier L, Zech W (2000) Persistence of soil organic matter in archaeological soils (*terra preta*) of the Brazilian Amazon region. In: Rees B, Ball B, Campbell C, Watson C (eds) Sustainable management of soil organic matter. CAB International, Wallingford, pp 190–194

Glaser B, Haumaier L, Guggenberger G, Zech W (2001) The terra preta phenomenon – a model for sustainable agriculture in the humid tropics. Naturwissenschaften 88:37–41

Goldberg ED (1985) Black carbon in the environment. Wiley, New York

Haumaier L, Zech W (1995) Black carbon – possible source of highly aromatic components of soil humic acids. Org Geochem 23:191–196

Hilbert P (1968) Archäologische Untersuchungen am mittleren Amazonas. Marburger Studien zur Völkerkunde, Berlin

Katzer F (1903). Grundzüge der Geologie des unteren Amazonasgebietes (des Staates Pará in Brasilien). Von Max Weg, Leipzig

Kögel I (1987) Organische Stoffgruppen in Waldhumusformen und ihr Verhalten während der Streuzersetzung und Humifizierung. Bayreuther Bodenkundl Berichte 1:131 pp

Kögel-Knabner I (1993) Biodegradation and humification processes in forest soils. In: Bollag JM, Stotzky G (eds) Soil biogeochemistry. Marcel Dekker, New York, pp 101–137

Petersen J, Neves EG, Heckenberger MJ (2001) Gift from the past: terra preta and prehistoric Amerindian occupation in Amazonia. In: McEwan C, Barreto C, Neves EG (eds) The unknown Amazon. British Museum, London, pp 87–105

Ranzani G, Kinjo T, Freire O (1962) Ocorrência de "Plaggen Epipedon" no Brasil. Bo Tecn Cient da Esc Sup da Agric "Luiz de Queirzo". University of Sao Paulo, Piracicaba

Ross SM (1993) Organic matter in tropical soils: current conditions, concerns and prospects for conservation. Prog Phys Geog 17:265–305

Schmidt MWI, Noack AG (2000) Black carbon in soils and sediments: analysis, distribution, implications, and current challenges. Global Biogeochem Cycles 14:777–793

Schmidt MWI, Skjemstad JO, Gehrt E, Kögel-Knabner I (1999) Charred organic carbon in German chernozemic soils. Eur J Soil Sci 50:351–365

Seiler W, Crutzen PJ (1980). Estimates of gross and net fluxes of carbon between the biosphere and the atmosphere from biomass burning. Clim Change 2:207–247

Skjemstad JO, Janik LJ, Head MJ, McClure SG (1993) High energy ultraviolet photo-oxidation: a novel technique for studying physically protected organic matter in clay-and silt-sized aggregates. J Soil Sci 44:485–499

Skjemstad JO, Clarke P, Taylor JA, Oades JM, McClure SG (1996) The chemistry and nature of protected carbon in soil. Aust J Soil Res 34:251–271

Smith HH (1879) Brazil: the Amazons and the coast. Scribner's Sons, New York

Smith NJH (1980) Anthrosols and human carrying capacity in Amazonia. Ann Assoc Am Geogr 70:553–566

Smith NJH (1999) The Amazon River forest: a natural history of plants, animals, and people. Oxford University Press, Oxford

Sombroek WG (1966) Amazon soils. A reconnaissance of the soils of the Brazilian Amazon region. Verslagen van Landbouwkundige Onderzoekingen, Wageningen, 283 pp

Sombroek WG, Nachtergaele FO, Hebel A (1993) Amounts, dynamics and sequestering of carbon in tropical and subtropical soils. Ambio 22:417–426

Tiessen H, Cuevas E, Chacon P (1994) The role of soil organic matter in sustaining soil fertility. Nature 371:783–785

WinklerPrins AMG (2002) Linking the rural with the urban: house-lot gardens in Santarém, Pará, Brazil. Urban Ecosyst (in press)

Woods WI, McCann JM (1999) The anthropogenic origin and persistence of Amazonian dark earths. Yearbook Conf Latin Am Geogr 25:7–14

Woods WI, McCann JM, Meyer DW (2000) Amazon dark earth analysis: state of knowledge and directions for future research. In: Schoolmaster FA, Clark C (eds) Papers and Proc Applied Geography Conf, Florida, pp 114–121

Zech W (1997) Tropen – Lebensraum der Zukunft. Geog Rundsch 49:11–17

Zech W, Pabst E, Bechtold G (1979) Analytische Kennzeichnung der Terra preta do indio. Mitt Dtsch Bodenkundl Ges 29:709–716

Zech W, Haumaier L, Hempfling R (1990) Ecological aspects of soil organic matter in tropical land use. In: McCarthy PM, Clapp CE, Malcolm RL, Bloom PR (eds) Humic substances in soil and crop sciences. Selected Readings. American Society of Agronomy and Soil Science Society of America, Madison, Wisconsin, pp 187–202

Zech W, Bäumler R, Sovoskul O, Sauer G (1996) Zur Problematik der pleistozänen und holozänen Vergletscherung Süd-Kamtschatkas – erste Ergebnisse bodengeographischer Untersuchungen. Eiszeitalter Gegenw 46:132–143

Zimmermann J (1958) Studien zur Anthropogeographie Amazoniens, Bonner Geographische Abhandlungen, Bonn

3 Evolution of the Scientific Knowledge Regarding Archaeological Black Earths of Amazonia

Dirse Clara Kern[1], Marcondes Lima da Costa[2], and Francisco Juvenal Lima Frazão[1]

3.1 Introduction

In Amazonia many small areas occur where soils were significantly affected by prehistoric man. These soils have a dark color, remains of archaeological materials, and higher Ca, Mg, Zn, Mn, P, and C contents in comparison with the adjacent soils. Known as archaeological black earth (ABE), Indian Black Earth, or *terra preta* (Kern and Kämpf 1989), these soils represent a remarkable example of how man can modify the original soil characteristics in a positive sense, improving its fertility. For this reason, they are frequently sought by the local people for subsistence cultivation of manioc, banana, maize, papaya, etc.

The ABE areas are found over the most diverse types of soil, including Latosol, Podsol, Podzolic, structured Purple Earth, and Petric Plintosols (Smith 1980; Kern 1988). They generally occupy 2–3 ha in area, but locally they can reach more than 80 ha (Hilbert 1955). Despite their frequency, ABEs are not shown on existing Amazonian soil maps because of the limited area that each comprises (Silva et al. 1970). The A horizon, which corresponds to the ABE layer, extends 40–60 cm deep on average, but in certain cases it can reach up to 2 m. It contains human occupation remains (ceramic fragments, lithic artifacts, and charcoal) through all of its thickness (Hartt 1885). Generally, ABEs are located on non-floodable ground (*terra firme*). The soils are well drained, often near rivers, creeks, or lakes, and almost always in a topographic position that permits a good view of the area as a whole. Nevertheless, on the Santarém Plateau, Nimuendaju (1948) found them far from the surface water sources, but rather having 2-m-diameter cylindrical wells. This author also mentions the presence of 1.5-m-wide roads in the forest, linking dark earth sites with each other.

[1] Museu Paraense Emílio Goeldi, Av. Perimetral n. 1901, Campus de Pesquisa, Belém, 66077-530, Pará, Brasil
[2] Centro de Geociências, Universidade Federal do Pará, Rod. Augusto Corrêa n. 1, Campus Guamá, 66075-900 Belém, Pará, Brasil

3.2
Hypotheses and Proposals for the Genesis
of Archaeological Black Earth Soils

Hartt (1885) suggested that the Indians were attracted to archaeological black earth (ABE) soils by their naturally high fertility. He based this hypothesis on the presence of ceramic fragments in all the occupational waste layers. For Camargo (unpubli. observ., 1941), Faria (1946), Cunha Franco (1962), and Falesi (1970, 1972), ABEs would have been originated from geological events or from the bottom of extinct lakes. According to Camargo, they would have their origin from volcanic ash, while Faria (1946) attributed their formation to sediments deposited at the bottom of extinct lakes or decomposition of volcanic rocks. For Cunha Franco (1962), ABEs would be associated with ancient lakes, on whose margins the Indians dwelt. Into these lakes they deposited the major part of the village waste, including many ceramic fragments. He based his hypothesis on the ABE spatial configuration (roughly circular) and their distribution (situated on dry ground, but not far from river margins). He assumed that ABEs near the rivers are due to the indigenous housing, but, even so, there should be evidence of lakes or creeks of former epochs. Falesi (1970, 1972) favored the thesis of Cunha Franco (1962) and suggested that these lakes would have been formed during Andean region uplift at the end of the Tertiary. At that time, according to the author, the rivers that flowed to the Pacific Ocean would have started to flow in the inverse direction, going to the Atlantic Ocean, and in the process damming much water in troughs. As these trapped water bodies ultimately dried out, the living organisms within them died, decomposed, and, consequently, enriched the soils at their base.

Ranzani et al. (1962) and Andrade (1986) classified the A horizon of the soil with ABE as being a "plaggen epipedon", that is, intentional incorporation of nutrient-rich material into the soil by management practices. Ranzani et al. (1962) attributed ABE fertility to the efficient use of the earth by the people of Andean origin who incorporated animal and vegetative ashes into the soil, with predominance of the former. Andrade (1986) provided three anthropogenic hypotheses for ABE formation from the Araracuara region in Colombia. These involve garbage waste, cultivation, or habitation. With the first two hypotheses the soil would have been intentionally enriched. The hypothesis that ABEs have been habitation places was discarded by Andrade (1986), because she found no indication of soil compaction in the Araracuara area of the Colombian Amazon region. However, the B horizons of ABE from the Oriximina region in Pará State, Brazil, are denser than those from the surroundings, even when they present equal contents of organic matter and clay, a fact that would suggest a compaction of this soil in the past. The high organic matter content in ABE, associated with deeper root penetration, could be responsible for the different soil structure developed over time (Kern 1988).

Through geological, pedological, and archaeological evidence, authors such as Gourou (1950), Hilbert (1955), Meggers and Evans (1957), Sombroek (1966), Simões (1982), Kern (1988), Kern and Kämpf (1989), Zech et al. (1990), Kern (1996), Kern and Costa (1997), Kern et al. (1999), and Glaser et al. (2001) suggested that ABEs would be places of former indigenous settlements, that is, they were formed through human occupation. Others (Woods and McCann 1999) reject explicitly the hypothesis of these soils having been formed only by accidental organic matter accumulation on top of the original soil by prehistoric settlements. They saw evidence that the soils were intentionally enriched to permit intensive semi-permanent agriculture.

Simões (1972) agrees that many prehistoric groups had their subsistence based upon grains and root cultivation, complemented with hunting, fishing, and collecting, permitting a more prolonged permanence in the place. However, from the ethnographic data, one verifies that the organic residues like the remains of food, leaves, seeds, and discarded fruit and vegetable skins generally are disorderly deposited in the surroundings and in some cases inside the habitations. Baldus (1942), studying the Kayapó tribe, stresses that "the Indians, in general, do not fear much the dirt, nor on their bodies nor in their houses nor their belongings. The major part of the trash, for this reason, is left where it falls, if it does not happen to be of interest to dogs and other animals that roam around and inside the houses." Roquette Pinto (1950) noted that the trash is randomly discarded around the habitations. "Single or in heaps, broken coconuts, jatobá broad beans, cobs of corn, charcoal, mixed with the rests of food and disabled ceramic utensils" were found.

The garbage of the people who inhabited the Amazon region must have been important for the increase in organic matter in the soil. A specific example is the palm tree, which had, and still has, several uses for the indigenous and mestizo communities. Between 200 and 250 different palm tree species occur in the region, of which 40 % is used by man in some way, be it for housing, nourishment, medicine, or ornamentation. Through ethnographic studies and reports of naturalist travelers, one knows that the forest people who inhabited (Indians) or who still inhabit (Indians and mestizos) the Amazon region built their houses with leaves of palm trees such as babaçu, buriti, bacaba, carana, etc. (Kern 1996). According to information regarding the mestizos who live in the Caxiuanã region, the houses with such leaf cover last, on average, only 3 years, resulting in large quantities of organic material remaining as waste (Kern 1996; Murrieta et al. 1989), studying the riverside community of Ponta de Pedras, Marajó Island, Pará State, mentioned an almost total use of the açai tree, of which the stem is used for construction and the leaf for the manufacture of *matapi* and *peconha* (an instrument to catch prawns and a type of tie bound to the feet to climb trees, respectively), while the fruit and the stem heart are used as food, either for man or for domestic animals. The results of chemical analyses of the palm tree leaves show elevated contents of Zn, Mn, P, Mg, Ca, and Na (Kern et al. 1999). The importance of palm trees in the everyday life of the present and past Amazon people and their specific chemical content does suggest that the palm tree residues contributed predominantly to the elevated content of some chemical elements found in ABE, principally those of Zn and Mn.

In addition to the garbage, funeral practices must have contributed to the increase in some chemical elements in ABE. At present, various tribal groups bury their dead inside their own houses or in the village center. The burial can be primary or secondary (Baldus 1942; Arnoud 1966; Oliveira 1968; Ramos 1971, 1980; Agostinho 1974; Gallois 1983). Some groups cremated their dead, the ashes being left at the place of cremation, or being drunk in a ritual. Others simply abandoned the house or the village, leaving the dead in the hammock, or put fire to the house (Migliazza 1964; Ramos 1971, 1980). The prehistoric groups dislocated and buried many times their dead in funeral urns which were deposited inside the village (Simões 1972). These cultural practices may be responsible for the morphological and chemical soil variations detected in the interior of archaeological sites.

3.3
General Morphological and Chemical Characteristics of ABE Sites in Amazonia

3.3.1
Morphological Characteristics

The A horizon of ABE, which corresponds to the human occupation layer, varies generally from 15–80 cm. The soil color, when humid, is black (7.5YR2/0). In comparison to adjacent areas, ABEs present a somewhat lighter texture, they are better structured (fluffier), and they generally contain a high density of ceramic fragments, which diminishes with depth. The thickness of the A horizon of the adjacent areas varies from 10–15 cm, its soil color is brighter, and the structure is normally weak. The thickness of the AB and BA transitional horizons in the ABE varies from 20–60 cm. They present darker colors than the adjacent areas, varying from black (10YR2/1) to a very dark brown (10YR2/2). They also show somewhat lighter (sandier) texture and are better structured than the adjacent areas. In the B horizons, the soil sections of the ABE and adjacent areas do not present, in general, large differences in their morphological characteristics.

3.3.2
Chemical Characteristics

A number of researchers have analyzed soil from ABE areas, mainly for Ca, Mg, P, and C (Cunha Franco 1962; Ranzani et al. 1962; Sombroek 1966; Falesi 1970, 1972; Silva et al. 1970; Vieira 1975; Bennema 1977; Zech et al. 1979; Smith 1980; Eden et al. 1984; Kern and Kämpf 1989; Pabst 1991; Glaser 1999;

Woods and McCann 1999). These studies have demonstrated the high fertility of these soils, standing out in relation to those commonly found in the region. In all analyzed samples, the Ca levels were more elevated than those of Mg, K, and Na, because of the major affinity with the cation exchange capacity of soil. A maximum value of $520\,mmol_c\,kg^{-1}$ was registered in the ABE of Belterra (Pabst 1991) and a minimum, below $10\,mmol_c\,kg^{-1}$ in the Colombian ABE (Eden 1984). The highest levels of Mg were detected in the ABE of Cachoeira-Porteira and Belterra regions, in the order of $70\,mmol_c\,kg^{-1}$ (Kern and Kämpf 1989; Pabst 1991), whereas the changeable P reached a maximum value of $1{,}500\,mg\,kg^{-1}$ in the Santarém region (Zech et al. 1979). The maximum organic matter content was found in the ABE of Belterra and achieved $210\,g\,kg^{-1}$ (Pabst 1991). For this author the organic matter in the ABE, besides differing in terms of quantity, also differs in quality: it is more stable and contains more organo-metallic compounds than those of the natural Amazonian Latosols. According to Glaser et al. (2001), the frequent charcoal findings and highly aromatic humic substances suggest that residues of incomplete combustion of organic material (black carbon) are a key factor for the persistence of soil organic matter in these soils. The levels of Ca, Mg, Zn, Mn, P, and C are higher in the A horizons, diminishing significantly with depth. The B horizons normally do not present significant variations in chemical aspects. This holds also for their physical and morphological aspects, proving that these soils were initially (before human occupation) not different from the adjacent soils.

The chemical analyses carried out with the ABE of Amazonia have shown that the archaeological site by itself is a great anomaly, but even so there are variations between and within the sites. Cunha Franco (1962) and Falesi (1970, 1972) showed the site area center as the thickest place where the sites had a roughly circular shape, like a lens buried with the plane part upward. Pabst (1985), besides identifying the area center of ABE as being the thickest place, also showed it to be the place where the highest levels of chemical elements were found. In the Cachoeira-Porteira and Oriximiná regions, where systematic investigations were carried out along north–south and east–west transects through three sites with ABE, one verified that the distribution of organic C, Ca, Mg, P, Zn, and Mn contents indicates preferential areas for waste deposition (Kern 1988; Ferreira and Kämpf 1989; Kern and Kämpf 1989, 1990, 1993). In the research carried out on ABE in the Oriximiná and Caxiuana regions, Pará State (Brazil), one verified that the Ca, Mg, Zn, Mn, and P contents presented places with major concentrations, differing from places with low values, but still significantly superior to those of soils of adjacent areas (Kern 1988, 1996; Kern and Kämpf 1989).

In a systematic geochemical survey carried out at one ABE site (Manduquinha) in the Caxiuana region, it was possible to reconstruct hypothetically the prehistoric settlement pattern (Fig. 3.1). The Indian group that inhabited this place did discard the material in specific and differentiated places. In the west portion of the site, the discarding of Mg-, P-, and Ca-enriched material

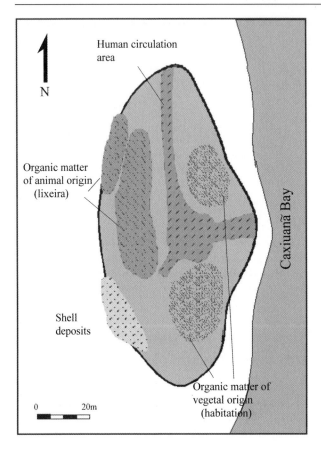

Fig. 3.1. Hypothetical reconstruction of the activity areas in the Manduquinha site, through interpretation of the geochemical data. (Kern 1996)

predominated which was associated with the refuse of food, mainly of animal origin, such as bones. According to ethnographic data, several groups that inhabited the region discarded the food refuse in the back part of their houses, where the kitchen was located. In this site one verified Zn, Mn, and Cu anomalies to the southeast and northeast, which can be related to vegetal organic matter used to cover the house walls and roofs, confirmed by ethnographic data. Places with relatively low levels of ABE-typical elements were identified: (1) at the central portion, a site that could have been a plaza and for this reason left intentionally cleaner; (2) the ABE northern edge adjacent to the jungle; and (3) the east part which provides access to Caxiuanã Bay, as the main source of water and fish for the group.

The studies carried out thus far on ABE sites allow the description of a evolutionary model for the development of these soils. The Manduquinha site, located in the Caxiuanã-Pará region, is discussed as an example. This site is located in the western side of Caxiuanã Bay, ca. 6 m above water level (Fig. 3.2). It was observed that before the occupation of the site there was already a Yellow Latosol, formed during the tropical pedogenesis that

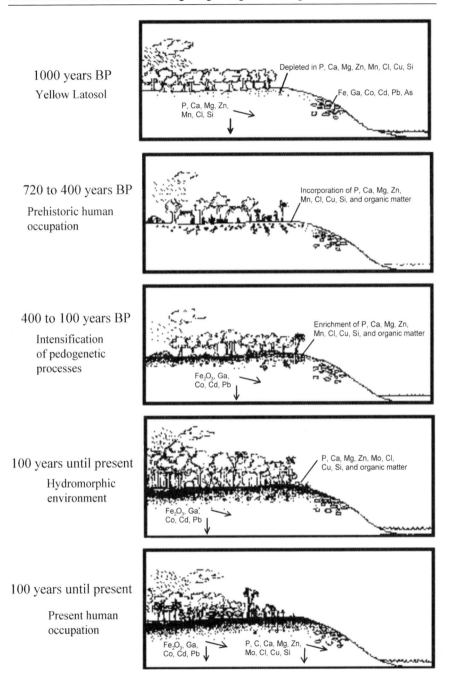

Fig. 3.2. Process of formation of Black Earth at the Manduquinha site. (Kern 1996)

occurred during the Quaternary. The source materials for this soil were the clayey, silty, and sandy sediments which correspond to the Alter do Chão formation of Cretaceous age, subsequently lateritized. These sediments were derived from sedimentary rocks and had been submitted to intense leaching; as a consequence, they are depleted in Ca, Mg, P, Zn, and Mn and show elevated levels of Fe and Al oxides and kaolinitic silicate clay minerals. The Latosols are well developed in the higher parts of the local landscape, coinciding with the places chosen for living by man in prehistoric times.

At least 720 years b.p., the Manduquinha site was settled and its inhabitants occupied the area for around 300 years. Throughout this period much organic material was incorporated into the soil. As already mentioned, ethnographic data show that the organic occupation residues were often randomly deposited in the surroundings and in some cases inside the habitations.

Kern (1996) supposed that due to the human occupation, the Manduquinha site was stage for an intense accumulation of organic material that overtook normal leaching processes (Fig. 3.2). Around a.d. 1600, the pedogenic processes intensified in an inverse direction to what had occurred previously during the site's inhabited phase. The discarded organic residue was transformed, releasing chemical elements that went into the soil complex. Ca, Mg, P, K, and other elemental additives could have been adsorbed to the cation exchange complexes, absorbed by plants, or leached. The added organic matter influenced directly the soil physical properties, favoring the formation of aggregates, reducing the cohesion and plasticity, and providing favorable conditions for aeration and tilth.

Therefore, the pedogenetic processes, acting intensely over abandoned areas with a large accumulation of organic material, allowed the elements cited above to be incorporated into the soil. After some time, an organic matter complexation may have occurred with ions of Ca and P and the clay particles (Sombroek 1966), consolidating ABE formation. At present, this and other ADE sites are used by the rainforest people for growing their subsistence crops or as habitation places.

3.4
Conclusions

The garbage of the people who lived in the Amazon region was very important for the increase in organic matter in the soil and also for its Ca, Mg, Zn Mn, P, and C enrichment. Waste products of vegetative and animal origin produced large quantities of organic matter that remained at the place of habitation. In addition, the vegetative matter when not used for food consumption or feed had and often still has multiple uses either by the indigenous population or by the present-day mestizo. Palms, in particular, are the predominant material used to cover houses by the forest people, and they are considered an important source of characteristic chemical elements (Zn, Cu, and Mn) for the soils (Kern et al. 1999). Burial practices could also have played a role in the increase in certain chemical elements in the soil, mainly

Ca and P, because the ethnographic and archaeological data show that several groups bury their dead inside their own houses, or in the village center (Migliazza 1964; Ramos 1971, 1980). Thus, garbage, burial practices, and vegetation material used for house construction contributed considerably to the increase in certain chemical elements in ABE.

References

Agostinho P (1974) Kuaríp. Mito e ritual no alto Xingu. Universidade de São Paulo, 221 pp

Andrade A (1986) Investigacion arqueológica dos antrosolos de Araracuara. Fundación de Investigaciones Arqueológicas Nacionales Banco de la República, Bogatá

Arnoud E (1966) Os índios Galibí do rio Oiapoque: Tradições e mudanças. Bol Mus Para Emílio Goeldi Ser Antropol 3:1–52

Baldus H (1942) Aldeia, casa, móveis e utensílios entre os índios do Brasil. Sociológica 4:157–172

Bennema J (1977) Soils. In: Bennema J (ed) Ecophisiology of tropical crops. Academic Press, New York, pp 29–55

Cunha Franco E (1962) As "Terras Pretas" do Planalto de Santarém. Rev Soc Agrôn Vet Pará 8:17–21

Eden MJ, Bray W, Herrera L, McEvan C (1984) Terra preta soils and their archaeological context in the Caquetá basin of southeast Colombia. Am Antiq 49:125–140

Falesi I (1970) Solos de Monte Alegre. Inst Pesqui Exp Agropec Norte 54:106–110

Falesi I (1972) O estudo atual dos conhecimentos sobre os solos da Amazônia brasileira. Inst Pesqui Exp Agropec Norte Bol Tec 54:17–31

Faria JB (1946) A Cerâmica da Tribo Uaboí dos rios Trombetas e Jamundá. Conselho Nacional de Proteção ao Índio, Rio de Janeiro, pp 5–42

Ferreira GC, Kämpf N (1989) Efeito da ação antrópica nas propriedades químicas de solo com Terra Preta de Índio. In: Proc I salão de Iniciação Científica, UGRGS, PROPESP, Porto Alegre, 6–10 Nov, Resumos

Gallois D (1983) A casa Waiãpi. In: Novaes SC (ed) Habitações Indígenas. Universidade de São Paulo, pp 147–168

Glaser B (1999) Eigenschaften und Stabilität des Humuskörpers der Indianerschwarzerden Amazoniens. Bayreuther Bodenkundl Ber 68:196 pp

Glaser B, Haumaier L, Guggenberger G, Zech W (2001) El fenómeno de Terra Preta – un modelo para una agricultura sostenible em los países tropicales. Grupo de trabalho: Terras Pretas Arqueológicas na Amazônia: o estado da arte. Anais de resumos SAB2001 – a arqueologia do novo milênio. Sociedade de Arqueologia Brasileira, Rio de Janeiro

Gourou P (1950) Observações geográficas na Amazônia. Rev Bras Geol IBGE 2:171–250

Hartt F (1885) Contribuição para a Ethnologia do Valle do Amazonas. Arch Mus Nac Rio Janeiro 6:10–14

Hilbert P (1955) A cerâmica arqueológica da região de Oriximiná. Instituto de Antropologia e Etnologia do Pará, Belém, 76 pp

Kern DC (1988) Caracterização Pedológica de solos com terra arqueológica na região de Oriximiná-PA. Tese (Mestrado em Solos), Curso de pós-graduação em Agronomia, Universidade Federal do Rio Grande do Sul, Porto Alegre, 231 pp

Kern DC (1996) Geoquímica e pedogeoquímica de sítios arqueológicos com Terra Preta na Floresta Nacional de Caxiuanã (Portel-Pará). Tese (doutorado em Geoquímica), Curso de pós-graduação em Petrologia e geoquímica, UFPa, Belém

Kern DC, Costa ML (1997) Solos Antrópicos de Caxiuanã. In: P.L.B. Lisboa – Caxiuanã. Museu Paraense Emílio Goeldi, Belém, pp 105–119

Kern DC, Kämpf N (1989) O Efeito de Antigos Assentamentos Indígenas na Formação de Solos com Terra Preta Arqueológica na Região de Oriximiná-Pa. Rev Bras Ci Solo Campinas 13:219–25

Kern DC, Kämpf N (1990) Características Físicas e Morfológicas dos Solos com TPA e sua importância para os estudos Arqueológicos. Rev CEPA 17(20):277-85

Kern DC, Kämpf N (1993) Antropogenic and pedogenic processes at Indian earth archaeological sites of Cachoeira-Porteira, Pará State, Brazil. In: Proc Resumos do Intl Symp on the Quaternary of Amazonia, ABEQUA, Manaus, Brazil

Kern DC, Frazão FJL, Costa ML, Frazão E, Jardim MA (1999) A influência das palmeiras como fonte de elementos químicos em sítios arqueológicos com Terra Preta. In: Proc 6th Simpósio de Geologia da Amazônia, Belém, Boletim de Resumos expandidos, SBG, Manaus

McCann JM, Woods WI, Meyer DW (1999) Organic matter and Anthrosols in Amazonia: interpreting the Amerindian legacy. In: Rees RM, Ball BC, Campbell CD, Watson CA (eds) Sustainable management of soil organic matter. CAB International, Wallingford, UK, pp 180-189

Meggers BJ, Evans C (1957) Archaeological investigations at the mouth of the Amazon. Bulletin 167. Bureau of Americam Ethnology, Smithsonian Institution, Washington, DC

Migliazza E (1964) Organização social dos Xiriâna do rio Uraricaá. Bol Mus Para Emílio Goeldi Ser Antropol 22:1-19

Murrieta BJ, Brondezio E, Siqueira A, Moran E (1989) Estratégia de Subsistência de uma População Ribeirinha do Rio Marajó-Açu, ilha do Marajó, Brasil. Bol Mus Para Emílio Goeldi Ser Antropol 5:147-163

Nimuendaju C (1948) Os Tapajós. Bol Mus Para Emílio Goeldi 10:93-106

Oliveira AE (1968) Os índios Juruna e sua cultura nos dias atuas. Bol Mus Para Emílio Goeldi Ser Antropol 35:1-28

Pabst E (1985) Terra Preta do Índio: Chemische Kennzeichnung und ökologische Bedeutung einer brasilianischen Indianerscharzerde. Master's Thesis, Ludwig-Maximilian-Universität, München

Pabst E (1991) Critérios de Distinção entre Terra Preta e Latossolo na Região de Belterra e os seus significados para a Discussão Pedogenética. Bol Mus Para Emílio Goeldi Ser Antropol 7:5-19

Ramos A (1971) As culturas indígenas. Livraria editora da casa do estudante do Brasil, Rio de Janeiro, Publ 2, 320 pp

Ramos A (1980) Hierarquia e simbiose. Hucitec, São Paulo, 246 pp

Ranzani G, Kinjo T, Freire O (1962) Ocorrência de "Plaggen Epipedon" no Brasil. Bol Tec Cient Esc Sup Agric "Luiz de Queiroz" 5:1-11

Roquette Pinto e (1950) Rondônia. Ser 5, vol 39. Brasiliana, São Paulo, 395 pp

Silva BN, Araujo JV, Rodrigues TE, Falesi IC, Reis RS (1970) Solos da área de Cacau Pirêra-Manacapuru. Inst Pesqui Exp Agropec Norte 2:1-198

Simões MF (1972) O Museu Goeldi e a Arqueologia da Bacia Amazônica. In: Roque Carlos – Antologia da Cultura Amazônica. Amazônia Edições Culturais, São Paulo, pp 172-180 (Antologia-Folclore, 6)

Simões MF (1982) A Pré-História da Bacia Amazônica: Uma tentativa de reconstituição. In: Cultura Indígena, textos e catálogo. Semana do Índio, Museu Goeldi, Belém, pp 5-21

Smith NJH (1980) Anthrosols and human carrying capacity in Amazônia. Ann Assoc Am Geogr 70:553-566

Sombroek W (1966) Amazon soils: a reconnaissance of the soils of the Brazilian Amazon region. Center for Agricultural Publications and Documentation, Wageningen, 292 pp

Vieira LS (1975) Manual da Ciência do Solo. Agronômica Ceres, São Paulo, 464 pp

Woods WI, McCann JM (1999) The anthropogenic origin and persistence of Amazonian Dark Earths. Yearbook Conf Latin Am Geogr 25:7-14

Zech W, Pabst E, Bechtold G (1979) Analytische Kennzeichnung vom Terra Preta do Índio. Mitt Dtsch Bodenkundl Ges 29:709-716

Zech W, Haumaier L, Hempfling R (1990) Ecological aspects of soil organic matter in tropical land use. In: McCarthy P, Clapp CE, Malcolm RL, Bloom RP (eds) Humic substances in soil and crop sciences. Selected Readings. American Society of Agronomy and Soil Science Society of America, Madison, Wisconsin, pp 187-202

4 A Geographical Method for Anthrosol Characterization in Amazonia: Contributions to Method and Human Ecological Theory

Laura A. German[1]

4.1
Introduction

This paper focuses on methodological and theoretical aspects of research to characterize the degree of pedological modification of *terra preta do índio* or Indian Black Earth, a class of anthrosols known in the recent literature as Amazonian dark earths (Woods and McCann 1999). Research was carried out along the Rio Negro, a blackwater region renowned for its oligotrophic (nutrient-poor) status and for the constraints these conditions place on the productivity of terrestrial, aquatic, and human ecosystems. For the blackwater *terra firme* environments where it is found, Black Earth represents an anomaly both ecologically and culturally. Its high nutrient and soil organic matter (SOM) contents contrast with the characteristics of the highly weathered soils that predominate in the region. Furthermore, both contemporary agricultural practices on Black Earth and the density and/or duration of settlements presumably required to form Black Earth are anomalous for theoretical models that draw strong linkages between specific ecological conditions, shifting agriculture, and low population density in *terra firme* environments.

The first contribution of this paper is methodological. A geographical method for characterizing Black Earth site distribution and the degree of pedological modification of these sites are presented. Interviews with local residents and systematic sampling procedures were employed to determine the spatial distribution of anthrosols along perennial waterways. Composite sampling, test pits, and total phosphorus determinations were used to determine the degree of soil modification at these sites. Possible applications of these geographical methods are discussed.

Additional contributions of research are theoretical. Analysis of the spatial patterns of human settlements permits the identification of preferred sites of Amerindian occupation and an improved understanding of the ability of these oligotrophic environments to sustain sedentary populations. Furthermore, the generation of indices of site development allows for important associations to be drawn between preferred areas of residence and the pattern of critical resources that may have sustained these inhabitants.

[1] World Agroforestry Centre (ICRAF), P.O. Box 26416 (ICRAF/AHI), Kampala, Uganda

4.2
Environmental Background

4.2.1
Amazonian Ecosystems

The typology of Amazonian landscapes and riverine systems is perhaps best presented in the seminal work by Sioli (1984). He divided Amazonian ecosystems into three major categories: whitewater, clearwater, and blackwater. This classification is based on limnology and on the characteristics of drainage basins that give rise to such divergent aquatic ecosystems. Whitewater ecosystems are those whose catchment basins originate in the Andes, where soil erosion causes water to be highly turbid and inundated areas to be bathed annually with a rich layer of sediments. These floodplain environments, known locally as *várzea*, are known to be the most fertile of the entire basin for agriculture and to have sustained dense populations and complex chiefdom societies in the late prehistoric era (Roosevelt 1980).

Clearwater basins are those that drain older geological formations of the Brazilian and Guiana shields to the south and north of the Amazon River, respectively (Sioli 1984). Clearwater lacks the sediment found in waters draining the Andes, due to the highly weathered state of Brazilian and Guiana shield formations. While these inundated environments are not particularly acid or limiting to agriculture, they lack the annual deposition of nutrients that provides a distinctive advantage to cultivation in the *várzea*.

Blackwater ecosystems may be best described by catchment basins that exist at the extreme end of a gradient reflecting relative nutrient poverty or oligotrophy – called *campina* or *caatinga* in Amazonian Brazil. The headwater regions of these blackwater ecosystems are characterized by scrubby or open forest vegetation and by extremely weathered Podzols (Salgado Vieira and Oliveira Filho 1962; Herrera 1985). Podzols are sandy soils with distinctive spodic horizons where organic compounds are leached into the subsoil. It is here that organic humic and fulvic acids leach into relatively stagnant bodies of surface and groundwater, leading to the characteristic dark and acid black water. This acidity, together with the lack of nutrient-rich sediment, causes the inundated islands and white sand beaches of the Rio Negro to be extremely acid and infertile, effectively restricting agriculture to the *terra firme*.

4.2.2
Human Ecology of Amazonia

An early generation of Amazonian scholars noted little evidence of complex societies in the Amazon Basin, and posited direct linkages between environmental constraints and human adaptability. It was understood that *terra firme* agriculture, for example, was limited to shifting cultivation systems due to the need to derive nutrients from the plant biomass through the burn in the absence of nutrient-rich soils (Meggers 1971). Other authors claimed that protein was the strongest limiting factor for human occupants, due to the low productivity of terrestrial and aquatic ecosystems and the absence of protein-rich crops (Meggers 1971; Gross 1975). These associations have been increasingly criticized, however, for minimizing environmental heterogeneity as well as the fine-tuning of human adaptations to environment.

Sioli's (1984) tripartite typology of Amazonian ecosystems was used by the next generation of Amazonianists to further differentiate these associations, and to demonstrate the fine-tuning of Amerindian land-use strategies and settlement patterns to a more diverse environment (Roosevelt 1980; Moran 1991; Chernela 1993). The unique biophysical properties of blackwater regions were shown to pose the strongest constraints on the productivity of human ecosystems and human adaptability (Moran 1991). First, the acidic and oligotrophic conditions of aquatic ecosystems are known to limit primary productivity, and therefore the biomass of aquatic food webs. While aquatic fauna is extremely diverse (Goulding et al. 1988), the total harvestable biomass is known to be more limited than in nutrient-rich whitewater environments. Second, these same aquatic conditions cause inundated landforms to be unfit for agriculture, as they are bathed annually by acid, nutrient-poor waters. As such, agriculture is limited primarily to *terra firme* environments. While much of the Amazonian *terra firme* is known to pose chemical limitations to agriculture due to very weathered soil horizons, the *campina* and *caatinga* environments common to blackwater regions are the most oligotrophic of *terra firme* ecosystems (Salgado Vieira and Oliveira Filho 1962).

The relationship between Amerindian population densities and distributions and environment is still hotly debated. Despite earlier suggestions to the contrary, there is mounting evidence that indigenous societies along major waterways attained significant complexity by the late prehistoric period (Roosevelt 1992), with the *várzea* sustaining dense populations with complex sociopolitical organization. Furthermore, recent archaeological evidence suggests that complex societies were present in whitewater, clearwater and blackwater environments alike (Heckenberger et al. 1999).

The blackwater site of Heckenberger and colleagues is located on the lower Rio Negro, however, where numerous archaeological sites from permanent occupations span the length of a canal connecting the Rio Negro with the Rio Solimões. This settlement distribution suggests that important trade routes existed between populations inhabiting whitewater and blackwater environments. This leaves in doubt whether isolated blackwater environments were capable of sustaining large, complex societies.

4.2.3
Amazonian Dark Earths

Amazonian dark earths, or simply Black Earth (*terra preta* by local designation), are anthrosols that are most likely formed through high-intensity nutrient deposition and burning in Amerindian settlements (Heckenberger et al. 1999; Woods and McCann 1999). A critical threshold level of culturally induced biotic activity and soil nutrient retention status must be reached to catalyze the new "black earth dynamic" (Pabst 1991; Woods and McCann 1999; McCann et al. 2001). This critical threshold appears to underlie the permanence of Black Earth in the absence of ongoing cultural amendments, and suggests that some form of more permanent settlement was necessary for these soils to form.

Nutrients tend to accumulate in human settlements as a result of the concentration of materials from the surrounding landscape, such as through the harvest of crops, fish, game, and forest products. As a nutrient that is more stable in soils (Woods 1975; Eidt 1977; van Raij 1991), soil phosphorus is one of the best indications of human settlement. The total amount of phosphorus accumulated in any archaeological site in relation to that found in unmodified soils is therefore a good indication of the net sum of nutrients added to the soil through time, and therefore, too, of the relative importance of the site to human settlement through time. If it were possible to determine with precision the total amount of phosphorus deposited in an archaeological site, this would doubtless be the best indicator of anthropogenic impacts on the soil at any given site. The degree of anthropogenic soil modification may in turn be used as an index of the relative importance of distinct landforms as settlement areas for the region's Amerindian occupants. This is because the extent of anthropogenic soil modifications resulting from nutrient deposition and other cultural activities in semi-permanent settlements can be assumed to be proportional to the number of individuals residing in these areas in space or time. This relative estimate of soil modification will of course result in only a gross assessment of the relative density or duration of settlements, and rests upon the assumption that similar patterns of deposition characterize distinct settlements in space and time. Yet it is one of the most direct measures we have for assessing the relative importance of distinct landforms and ecosystems for human settlement in a region where the archaeological record (in particular, organic remains) is easily erased by intensive weathering.

The purpose of this research was to develop a relative measure of settlement intensity in blackwater regions through spatial survey and pedochemical characterization of anthrosols. In doing so, it allows us to more directly observe the relationship between environmental heterogeneity and the spatial pattern of settlements in these blackwater regions and elsewhere. The spatial differentiation of Black Earth into the darker, more nutrient-rich *terra preta* and the lighter *terra mulata* (Brown Soil) (Woods and McCann 1999) complicates this procedure due to the differentiation of formation processes.

However, only *terra preta* (herein designated "Black Earth") has the high levels of soil phosphorus that facilitate the determination of cumulative nutrient deposition or "anthropogenic impact." As such, measurements of soil phosphorus will specifically target these areas of prior Amerindian residence.

4.2.4
The Rio Negro

Research was carried out on the Rio Negro, a blackwater basin that extends from the central Amazon near Manaus to the northwestern reaches of Amazonian Brazil near the borders of Colombia and Venezuela (Fig. 4.1). Intensive research focused on the Lower Rio Negro, between the municipalities of Novo Airão and Barcelos. Countless Indian Black Earth sites of various sizes today dot the banks of the Lower Rio Negro, indicating that semi-permanent settlement or frequent reoccupation was common in earlier periods. Ethnohistorical accounts provide some indication as to the origin of these anthrosols.

4.2.5
The History of Human Occupation on the Rio Negro

The entire Rio Negro watershed, from the mouth to the headwaters, was territory of Arawak tribes from the North Maipuré linguistic family. This occupation spanned 'the time when proto-Arawak expanded up the Rio Negro until European penetration in the 18th Century' (Wright 1992). This is confirmed by ethnohistorical accounts which indicate that Manao Indians occupied much of the Rio Negro during early historic times (Ferreira Reis 1906; Porro 1987).

Other linguistic groups to occupy this blackwater region included Tukano and Maku linguistic families, which today occupy the upper reaches of the Rio Negro and the Rio Uaupés. Nimuendaju (1955) suggested that the contemporary ethnic composition of this northwestern region was formed through three strata: (1) the oldest, diverse ethnicities of semi-nomadic hunters and gatherers (Maku, Uaicá, and Xiriana); (2) populations of more advanced cultures at the beginning of the Christian era (Arawak and Tukano); and (3) the most recent "hybrid cultures" arising out of European contact with these native groups. Ethnographic research and oral tradition of contemporary Arawak and Tukano groups each suggest that groups from the Arawak linguistic family already inhabited the region when the Tukano arrived (Reichel-Dolmatoff 1985; Wright 1992).

Certain patterns of complex social organization today characterize the Arawak and Tukano groups of the upper Rio Negro region, including a system of social organization in which sibs are organized according to a hierarchy of ritualized roles and associated with certain territories and resources (Chernela 1993). Given that these cultural patterns are not shared by Tukano

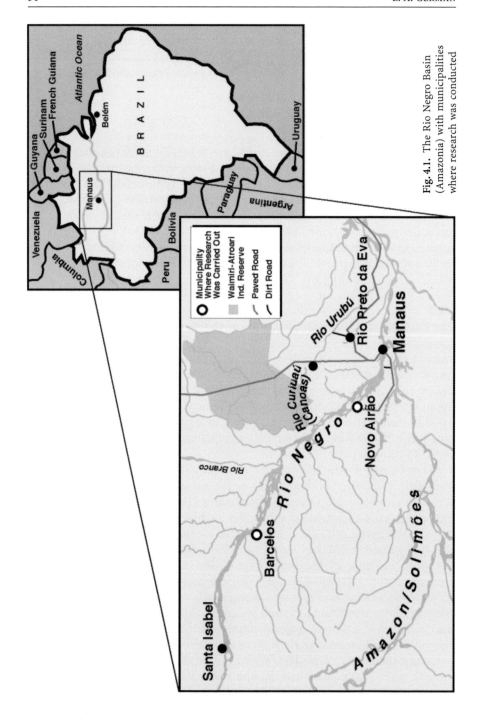

Fig. 4.1. The Rio Negro Basin (Amazonia) with municipalities where research was conducted

groups of western Amazonia (in their supposed region of origin), it is likely that the ritual complex and system of social organization were assimilated by the Tukano from the Arawak (Wright 1992). The first ethnohistorical accounts about the Arawak (such as the Manao) indicate similar patterns of complex socio-political organization (Ferreira Reis 1906; Porro 1992; Wright 1992). Tukano ethnohistorical accounts suggesting sedentary horticultural activity among these groups are also curious for a region in which the predominance of Podzols limits Latosol distributions and also agricultural productivity (Salgado Vieira and Salgado Filho 1962; Klinge 1967; Salgado Vieira 1988). Arawak groups inhabiting the lower Rio Negro, where Latosols predominate, would certainly have been capable of sedentary occupation based on horticulture, fishing, and other subsidiary activities. These accounts would suggest that complex patterns of socio-political organization did exist, which in turn points to the likelihood of more permanent patterns of settlement.

This more sedentary occupation is likely to underlie the formation of Black Earth sites along the Rio Negro, given observations by Heckenberger (1999) that contemporary groups produce no extensive Black Earth deposits even after some 50 continuous years of occupation. These anthropogenic environments are likely an artifact of the more sedentary, complex societies inhabiting the Rio Negro during late prehistoric and early historic periods. Initial occupation dates associated with modified soil strata in sites on the Lower Rio Negro span a period between 6,850 years B.P. (on what ultimately became the largest and structurally most elaborate site) and 400 years B.P. Most of the smaller sites in this region were occupied no earlier than 1,125 years B.P. (Heckenberger et al. 1999). An assessment of the relative size and degree of soil modification of these sites will tie these ethnohistorical accounts into the pattern of human occupation at the landscape level and will help to reconstruct the human ecology of the region's past.

4.3
Material and Methods

4.3.1
Geographic Area

Research was carried out on the lower portion of the Rio Negro, between the municipalities of Novo Airão and Barcelos. This region was selected on the basis of accessibility and environmental heterogeneity. The latter was determined by the 1978 geological, topographical, phytoecological, and pedological maps of Radam Brasil, in which the stretch from the mouth of the Igarapé Grande (in the municipality of Novo Airão) to the mouth of the Rio Branco is shown to be highly heterogeneous for the Rio Negro Basin.

The sampling protocol was determined on the basis of research objectives: to determine the distributions of anthropogenic soil on the landscape, and to sample all of the possible manifestations of this anthropogenic soil modification. As such, rather than have known or documented Black Earth sites dictate the sampling procedure, river margins had to be researched exhaustively for the presence or absence of Black Earth and for its pedological properties. For this to be feasible for such a large region, the Lower Rio Negro had to be sampled. Exhaustive geomorphological research was therefore carried out on three 10-mile stretches of riverbank on the right margin of the Rio Negro, as illustrated by white lines within boxed areas in Fig. 4.2. Within each setting, the margin of the central channel of the Negro River and all corresponding tributaries were researched to gather data from diverse physiographic zones.

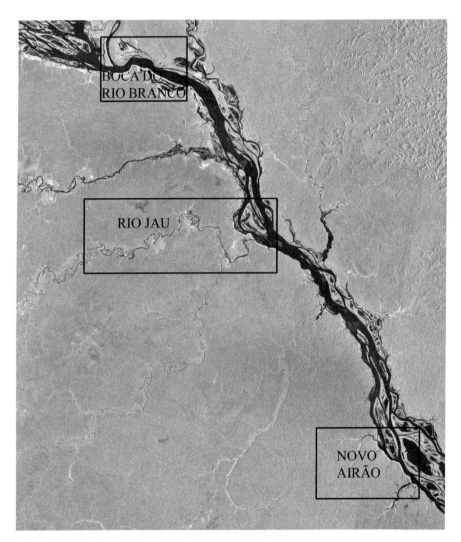

Fig. 4.2. Lower Rio Negro, with sites of intensive sampling

The three settings include: (1) "Novo Airão," a 10-mile stretch of the Rio Negro between the municipal center of Novo Airão and the *Igarapé do Sobrado* (a medium-sized tributary); (2) "Rio Jaú," a 10-mile stretch of the Rio Negro encompassing the mouth of the Jaú River; and (3) "Boca do Rio Branco," a 10-mile stretch opposite the mouth of the Branco River. Each of these areas encompasses 10 miles of the main channel of the Rio Negro and all perennial tributaries that fall within this area, including several *igarapés* (roughly, streams) of varying sizes in the Novo Airão setting, one large river for the Rio Jaú, and one minor tributary for the Boca do Rio Branco. It should be noted that many *igarapés* are actually much larger than streams. Their defining attributes are their small catchment basins relative to rivers, and the influence of larger catchments (those of the rivers into which they drain) on seasonal fluctuations in water level.

4.3.2
Identification, Characterization, and Classification of Black Earth Sites

To classify the relative extent of anthropogenic soil modifications in each setting, it was necessary to locate Black Earth sites, to characterize these sites with respect to their spatial, morphological, and pedochemical properties, and to determine the relative extent of anthropogenic soil modification for each site. Several strategies were employed to locate Black Earth sites: interviews with local residents, literature reviews, and random sampling on all areas of *terra firme* abutting active river channels. These sites were registered with the use of a GPS, and their anthropogenic origins confirmed through laboratory determination of total phosphorus. While anthrosols are defined by Spaargaren (1994) as having a minimum of 250 ppm total phosphorus, this is an arbitrary definition that fails to take into account diverse degrees and types of anthropogenic soil modification (see Woods and McCann 1999). For this research, classification of soils as "Black Earth" was done according to statistical differences in total phosphorus content in relation to control samples, for which there was no such cut-off.

After locating individual Black Earth sites, these sites were characterized morphologically and chemically. To determine the relative importance of diverse settlement areas, biophysical indices that best reflect the cumulative anthropogenic impact or historical importance of these sites had to be selected. This was done according to the literature, and what we know about the relationship between settlement duration and/or density on the one hand, and phosphorus deposition, soil coloration, and site size on the other. The following variables were selected for analysis:

4.3.2.1
Site Size

The approximate size of each Black Earth site was determined through field observations (soil color, vegetation characteristics, etc.) and with the assistance of local residents. An exact determination of the spatial extent of anthropogenic impacts was impossible due to the spatial heterogeneity of impacts, the often gradual pedological transition to non-anthropogenic soils and the limited time that could be dedicated to any given site when analyzing a large geographical area. This limitation was dealt with methodologically during the analytical stage, in which diverse indices jointly contribute to an assessment of site class.

4.3.2.2
Total Phosphorus

The difficulties determining site size also hindered making accurate assessments of the total amount of phosphorus accumulated in each site over time. Several sampling procedures were therefore utilized to generate a more robust assessment of total phosphorus content: (1) composite samples of randomly selected points throughout each Black Earth site, at a depth of 50 cm (this depth was determined to be the depth within which most anthropogenic impacts are concentrated for Black Earth sites of the Lower Negro. While the anthropic epipedon of a few sites extends beyond this depth, this is by far the exception. Extending the sampling depth to 1+ m would be necessary for some regions with dense Amerindian populations, yet would require additional resources for fieldwork); (2) one sample of the epipedon in the deepest and darkest part of the site (the "center"), where anthropogenic soil modification was most intense for that site; (3) one sample of each successive buried horizon from site centers (to 1 m); and (4) control samples at a depth of 50 cm, to decipher between soil P resulting from anthropogenic influences and that occurring naturally. These samples were later analyzed through digestion with concentrated nitric acid (for 5 h) and standard chlorimetric determination.

4.3.2.3
Pedological Analysis

At each site center, a qualitative assessment of the site's morphology was carried out to a depth of 1 m. Each horizon was analyzed for depth, color (the 1994 edition of the Munsell Soil Color Chart), and texture (rapid field assessment method). These assessments helped to determine the extent of color modification resulting from burning or other human activities, the depth of anthropogenic impacts, and patterns of phosphorus movement through the pedon (as a function of texture, use, or other factors). These indices do not always vary proportionately to one another, and no single index is both best representative of anthropogenic impacts and easily determined with precision. As such, these indices had to be combined so that the relative degree of site modification (used here as an index of the overall importance of each site for human settlement) could be determined for each site. For this purpose, a multidimensional scaling exercise was carried out such that diverse variables would contribute to a site's classification. Six variables were selected as indicative of relative anthropogenic impacts: depth of the anthropic epipedon at the site center, color (by value, the numerical color parameter most reflective of differences in anthropogenic influences), and four indices of total phosphorus.

Two indices of total P were generated from the composite sample (0–50 cm) and two from samples taken at the site center. The first two correspond to the total weight of P (in grams) found in: (1) the composite sample (for the volume of the auger); and (2) the Black Earth site (in which the value for the composite sample was extrapolated out across the estimated site area). The two indices of total P that were generated from the site center represent the total weight of P in the auger when sampling: (1) the anthropic epipedon and (2) the full pedon to 1 m. Due to preliminary assumptions that P would be highly concentrated in the epipedon, direct measurements of P in buried horizons were carried out for some sites only. For the remaining sites, P values for buried horizons were extrapolated from the epipedon on the basis of average distributions of P throughout the pedon in comparable sites. The influence of soil texture and use (tillage) on P migration down through the pedon was estimated and taken into account in these extrapolations (see below).

4.3.3
Correlating Site Class with Access to Critical Resources

The relationship between the location, degree of modification, and access to critical resources for each site was assessed qualitatively. Black Earth sites were first superimposed on Landsat images. These images were then classified spectrally in PhotoShop, using RADAM-Brasil maps and ethnographic data to "ground-truth" spectral signatures from satellite images. This permitted the spatial delineation vegetation classes, water types, and other landscape elements important to regional livelihood (lakes, beaches, etc.). While these images are limited to large-scale phenomena, and therefore useful for elucidating human adaptive behavior at a coarse resolution only, they help to demonstrate how the size and location of Black Earth sites co-vary with ecotones and with large landscape classes. These landscape classes are often, in turn, associated with specific adaptive strategies. Ethnographic research assisted in identifying critical resources for contemporary inhabitants and interpreting the spatial relationship between site size and distribution and the distribution of these resources. Together, these data permit visual inspection of Landsat images and site class overlays as a function of specific productive activities, ecological zones, and other possible factors (political-economic, health, aesthetic, etc.) leading to the preference of certain sites of residence over others.

4.4
Results

4.4.1
Characterization and Classification of Black Earth Sites

A total of 47 Black Earth sites were identified within sampled areas of the Lower Rio Negro: 17 in the vicinity of Novo Airão; 25 in Rio Jaú; and 4 in Boca do Rio Branco. While the low number of sites in the Rio Branco setting is in marked contrast to the other two settings, it is important to keep in mind the sharply divergent physiographies that characterize each setting. The number of tributaries adjoining the Rio Negro within the 10-mile stretch of river margin influences both the abundance of river bluff suitable to settlement and the total distance of riverbank sampled during fieldwork. While Novo Airão had a larger number of tributaries than the Rio Jaú and Rio Branco stretches, the Jaú River extends a much greater distance inland than any other tributary – permitting the sampling of an additional 97 km of riverbank (an arbitrary cut-off for a river that extends more than 200 km inland). Due to these discrepancies, the comparative analysis is not designed to contrast distinct sections of the Rio Negro for their relative importance to human settlement, but to pool settlement phenomena to assess those factors influencing settlement distributions for the region at large. Values for total phosphorus (P1–P3), epipedon depth, and color are given in Table 4.1 for each of these sites.

To generate the third index of total P (the weight of P in the auger to a depth of 1 m in the site center, P3), direct measurements were made for some sites and extrapolations from P content in the epipedon for others, as mentioned above. Despite preliminary assumptions that phosphorus – bound tightly to Fe and Al oxides – would be highly concentrated in the epipedon, this was tested for a number of sites through measurement of P distribution throughout the pedon (to 1 m). Contrary to what was expected, significant amounts of P were found in buried horizons – primarily in the B horizon, where significant pedoturbation had led to mixing with the epipedon. This suggests that future applications of this methodology should include full sampling of the site center to 1 m, or, minimally, of A and B horizons.

Sites for which sampling was carried out for the full pedon were differentiated according to their use histories and soil texture, in order to determine the influence of each of these on downward migration of phosphorus. From Table 4.2, it is evident that while no significant difference in the vertical distribution of phosphorus results from soil textural differences, significant

Table 4.1. Black Earth sites and indices of site modification

Site no.	Site name	P1 (g auger, composite)	P2 (g in H1)	P3 (g auger, center)	P4 (g in site)	Depth (cm)	Color value
1	Ardale	0.128	0.028	0.065	382	0.21	3
2	Baiano	0.602	0.707	1.106	3,594	0.45	3
3	Burití	0.620	1.164	1.447	2,465	0.08	2
4	Castanha	1.164	0.687	1.343	1,735	0.35	3
5	Dinheiro	0.816	0.901	0.836	4,868	0.50	3
6	Espíritu Santo	0.138	0.190	0.404	547	0.13	3
7	Etelvina	0.125	0.256	0.352	248	0.45	3
8	Jamal	1.193	1.321	2.498	10,683	0.55	3
9	Mário-Jorge	0.093	0.361	0.486	649	0.40	3
10	Miguel	1.440	0.116	0.211	25,782	0.44	3
11	Nova Esperança	0.033	0.172	0.369	198	0.30	3
12	Santana	1.355	1.459	1.874	619	0.50	5
13	Vista Alegre-B	0.896	1.222	1.886	13,372	0.55	3
14	Vista Alegre-C	0.877	2.116	3.242	5,236	0.48	3
15	Art-Rup – Boca	1.309	0.755	1.498	11,716	0.35	3
16	Art-Rup 2 – Pedral	5.813	2.173	2.795	46,262	0.35	3
17	Art-Rup2 – Boca	3.705	0.362	0.605	14,741	0.22	3
18	Ataíde	0.053	0.056	0.211	210	0.55	3
19	Boca Jaú-2A	0.522	0.535	0.392	1,453	0.30	3
20	Boca Jaú-2B	0.585	0.202	0.435	2,676	0.30	3
21	Cachoeira II	0.173	0.065	0.193	1,375	0.20	3
22	Cachoeira III	0.713	1.138	1.212	7,096	0.63	4
23	Délio I	0.408	0.387	0.457	811	0.25	3
24	Erasmo	0.223	0.239	0.673	889	0.22	3
25	Heck-3	1.412	0.643	0.989	5,618	0.20	2
26	Miratucú	1.554	1.091	2.399	27,829	0.22	2
27	Nazaré	0.543	0.219	0.445	2,699	0.28	2
28	Pajé	1.165	0.884	1.178	4,637	0.18	3
29	Pedral	0.024	0.596	0.755	29	0.56	3
30	Ponta Pedra-Negro	2.497	2.479	3.428	9,937	0.40	3
31	Quebrado	0.400	0.234	0.279	398	0.25	3
32	Quebrado II	0.185	0.294	0.420	184	0.35	4
33	Velho Airão	7.306	6.581	7.379	13,0804	0.40	3
34	Vista Alegre	1.244	1.064	1.031	7,427	0.33	3
35	Volta	0.410	0.203	0.427	1,631	0.11	3
36	Zé Ribeiro	0.199	0.298	0.508	816	0.50	4
37	Zé-RB2	–	0.180	0.274	331	0.13	3
38	Pedro Barata	0.609	0.448	0.485	1,091	0.29	3
39	Dona Cota	0.600	0.707	0.757	382	0.68	3
40	Sítio Sr. Mário	1.223	3.774	4.096	73	0.71	3
41	Sítio Mulata	0.590	2.128	2.325	10,570	0.56	4
42	Gumercindo	0.002	0.021	0.039	8	0.57	3
43	Nonato	0.545	2.125	2.302	5,377	0.46	3
44	Heck-5	1.412	0.381	0.637	5,618	0.14	3
45	Cap. Grande	0.182	0.664	1.022	1,088	0.20	3
46	Waldemar	0.322	0.357	0.64	1,604	0.42	3
47	Jaú 2-C	0.382	0.125	0.539	2,283	0.30	2

Table 4.2. Downward migration of total phosphorus as a function of texture and tillage

Horizon	Mean percentage of P in Hx: (1) As a function of texture		(2) As a function of use	
	Clay soils	Sandy soils	Max. tillage	Min. tillage
Mean, H1	76	78.6	52.9	79.1
Mean, H2	22.6	19.5	43.3	20.4
Mean, H3	1.4	1.9	3.8	0.5
Mean, H4	0	0	0	0

impacts result from soil tillage. These use histories and distribution patterns were utilized to extrapolate vertical P distributions on those sites for which samples from the site center were only taken from the epipedon. This procedure would, however, be eliminated should the recommended direct sampling of each horizon be carried out.

Several Euclidean multidimensional scaling operations were carried out to test the proximity of the 47 Black Earth sites in multidimensional space according to distinct combinations of variables (indices of site modification) from Table 4.1. This permitted classification of sites according to multiple properties that are each reflective of density and/or duration of past human settlements.

Black Earth sites are plotted in Fig. 4.3 according to their relationship to one another in multidimensional space. The removal of one or more indices of site modification had no significant effect on the relative placement of Black Earth sites on the results graph, indicating that variables tend to vary together and that the selected variables are, in fact, good indices of anthro-

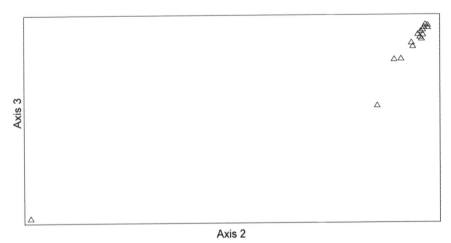

Fig. 4.3. Multidimensional scaling of Black Earth sites according to multiple indices of anthropogenic soil modification

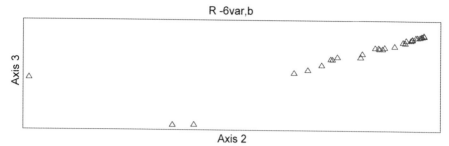

Fig. 4.4. Multidimensional scaling of Black Earth sites according to multiple indices of anthropogenic soil modification (outlier removed)

pogenic soil modifications. In Fig. 4.3 it can be seen that one site, Velho Airão, is the most modified site of those researched and a strong outlier.

The removal of this outlier gives a better representation of the relative degree of anthropogenic soil modifications of the remaining sites, which is represented in Fig. 4.4. This plot illustrates how sites tend to differentiate from one another along a single orientation in multidimensional space, indicating that sites differ by degree rather than by kind. While class I Black Earth sites could arguably include only the four most modified sites (Velho Airão, Arte Rupestre 2 – Pedral, Miratucú, and Miguel), sites were divided into three classes of equal size to better differentiate low-end sites. This decision may change according to the particular questions being addressed, or as this study is expanded to include intra-regional comparisons.

Table 4.3. Black Earth sites grouped by site class

Class I	Class II	Class III
Art-Rup (Boca 3)	Baiano	Ardale
Art-Rup 2-Pedral	Boca Jaú-2A	Ataíde
Art-Rup2 (Boca 3)	Boca Jaú-2B	Cachoeira II
Cachoeira III	Burití	Erasmo
Espíritu Santo	Cap. Grande	Etelvina
Heck-3	Castanha	Gumercindo
Jamal	Délio I	Heck-5
Miguel	Dinheiro	Mário-Jorge
Miratucú	Dona Cota	Nova Esperança
Pajé	Jaú 2-C	Pedral
Ponta Pedra-Negro	Nazaré	Quebrado
Sítio Mulata	Nonato	Quebrado II
Sítio Sr. Mário	Pedro Barata	Vista Alegre-C
Velho Airão	Santana	Volta
Vista Alegre	Waldemar	Zé Ribeiro
Vista Alegre-B		Zé-RB2

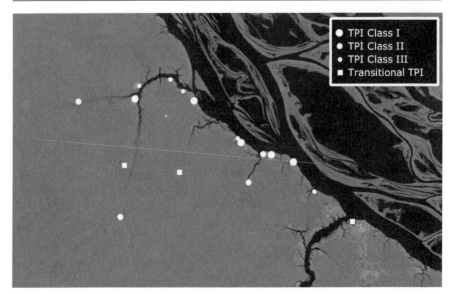

Fig. 4.5. Landsat image of the Novo Airão trajectory, with superimposed Black Earth sites

The resulting classification of Black Earth sites into classes I, II, and III is presented in Table 4.3, in which class I sites represent those subject to the highest degree of anthropogenic modification.

These sites have been plotted on Landsat images in each of the three locales: Novo Airão (Fig. 4.5), Rio Jaú (Fig. 4.6), and Boca do Rio Branco (Fig. 4.7). On these images, site class is represented by the size of white or black circles, with larger circles representing those sites classified as having

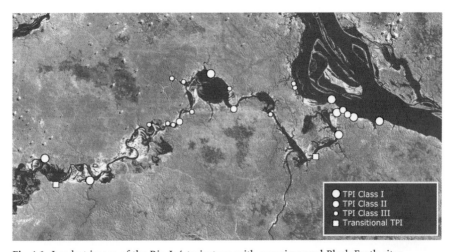

Fig. 4.6. Landsat image of the Rio Jaú trajectory, with superimposed Black Earth sites

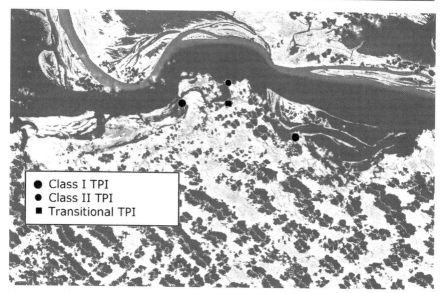

Class I TPI
Class II TPI
Transitional TPI

Fig. 4.7. Landsat image of the Boca do Rio Branco trajectory, with superimposed Black Earth sites

undergone a higher degree of anthropogenic modification. Sites that were transitional between classes are indicated in italics in Table 4.3, and by white or black squares on the figures.

4.4.2
Assessing Relative Access to Critical Resources

Analysis of the relationship between the location and degree of anthropogenic modification of Black Earth in relation to major landscape features is facilitated by the simplification of satellite images according to a few basic landscape classes. IBGE resource maps, spectral signatures of Landsat images, and accounts by contemporary residents of important resources found in diverse terrestrial and riverine environments permit this classification of landscape into basic categories. Spatial delineation of the landscape into *terra firme* forest, *igapó* (flooded forest), whitewater rivers, *várzea*, blackwater rivers, lakes, and beaches has been done for each of the three trajectories in Figs. 4.8, 4.9, and 4.10.

From a visual analysis of these maps, several observations can be made regarding large-scale patterns of site-resource associations, namely, Black Earth site location and class vis-à-vis major landscape units. First, Black Earth sites tend to be located on *terra firme* bluffs abutting active river channels, in particular where secondary tributaries meet the main channel of the Rio Negro or where two smaller tributaries meet, and in the headwaters of secondary streams (*igarapés*).

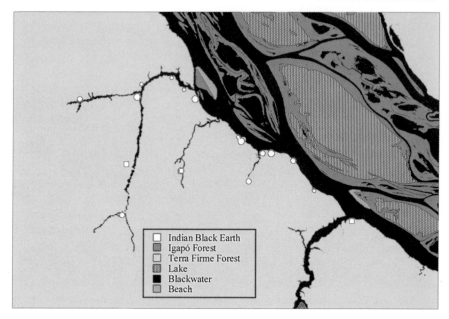

Fig. 4.8. Major landscape classes, Novo Airão trajectory

The orientation toward bluff settlement was documented by Denevan and Padoch (1987), and may be explained as a strategy that maximizes access to multiple ecological zones, while at the same time providing for strategic defense and perhaps also political-economic control over important resources. The latter would be especially true for sites located at the conflu-

Fig. 4.9. Major landscape classes, Rio Jaú trajectory

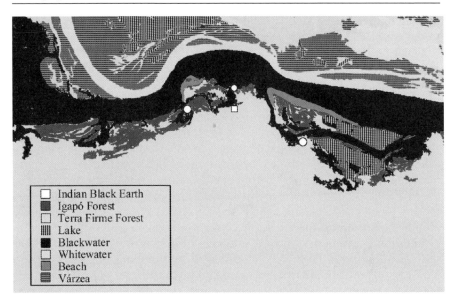

Fig. 4.10. Major landscape classes, Boca do Rio Branco trajectory

ence of two rivers, where products from far inland might be gathered by politically dominant groups at the river's mouth. The most significant class I site, Velho Airão (located on the far right of Fig. 4.9), was known to be occupied during the historical period by indigenous communities, and later by a Portuguese colony. Soil chemistry was sharply impacted by the sedentary occupation of these groups, who benefited economically from the spatial concentration of extractive products from the Jaú River Basin.

The most modified (classes I and II) bluff sites tend to be near lakes, whitewater, and the main channel of the Rio Negro, each area having distinctive advantages for resource procurement during certain seasons. Bluff settlement along main river channels provided access to flooded forest, where most fishing is carried out during the rainy season, and in lakes and the main river channel, where fish is concentrated in lower volumes of water during the dry season and thus easier to catch. The large Black Earth sites near the mouth of the Rio Branco also allowed access to whitewater regions, where higher primary productivity increased harvestable fish biomass. Bluff settlement on main river channels also provided access to beaches, where an abundance of turtle meat may have been found at the beginning of the dry season when nesting grounds were established along the white sand beaches of the Negro and Jaú Rivers.

The second tendency, in which settlements were located at the confluence of two rivers or *igarapés*, is logical for both livelihood and strategic purposes. These localities provided access to resources from a broader range of environments, strategic control over upstream resources, and visibility for defense. Furthermore, settlement locations near streams or *igarapés* provided access to limited terrestrial fauna during the dry season, when they descend to the river's edge for water and are easier to capture.

Finally, sites located in headwater regions were likely to have provided strategic access to terrestrial game populations, given the larger area of contiguous rainforest and the limited availability of aquatic protein in these areas. While settlements located along primary river channels ensure access to fish protein during most seasons, seasonal or semi-permanent settlement in the headwaters maximized access to terrestrial game and played an important role in alleviating seasonal shortages in fish protein. It is possible that a functional relationship existed between these headwater sites and those located at the mouth of these *igarapés*, either through seasonal migration or the exchange of goods. These relationships based upon the exchange of riverine and forest products have been noted for settled agriculturalists and more mobile hunters and gatherers in the Upper Rio Negro region (Ribeiro 1995).

Additional factors nevertheless influenced the selection of bluff and other settlement locations. From the above images, most notably Fig. 4.9 of the Rio Jaú, it is evident that settlements tended to be located near *igapó* forest, the seasonally flooded areas unfit for agriculture. Here, fallen fruits were a critical resource for sustaining fish populations (Goulding et al. 1988) and provided refuge from predation during the rainy season, creating an important fishing zone. Yet the infertility of *igapó* soils required that areas of contiguous *terra firme* be an important factor in site selection among horticulturalists. These factors are manifested in site distributions, which show Black Earth sites to be located on *terra firme*, near *igapó*, yet away from the larger, spatially contiguous areas of *igapó* where fish easily escape predation and capture.

4.5
Conclusions

Two objectives of this chapter were to generate and test a methodology for the geographical characterization of Black Earth sites, and to contribute to human ecological theory of Black Earth and blackwater ecosystems. The development of geographical methods for Black Earth site characterization helped to identify promising pedological indices of cumulative human impacts on the soil, which were used here as a proxy for the relative importance of diverse archeological sites for human settlement. This procedure oversimplifies historical process by assuming that similar settlement and land-use patterns characterized distinct sites and occupation periods. As an approach that is a yet little explored attempt to quantify human impacts directly through the measurement of total nutrient deposition, it merits further consideration and refinement. The process of identifying classes of Black Earth on the basis of distinctive formation processes, undertaken recently by Woods and McCann (1999), will contribute a great deal to the refinement of these methods.

The multidimensional scaling procedure to generate a site class index (degree of anthropogenic soil modification) generated more robust results than a priori selection of any single indicator of human impact. However, it is suggested that future applications of this methodology rely on more robust in-field sampling. This would eliminate the need to combine multiple indicators in the procedure to classify the degree of anthropogenic soil modification for each site, yet would rely on more extensive in-field sampling (i.e., of all buried horizons where significant amounts of P are found, and for more reliable assessments of site size). The impact of tillage on the downward migration of phosphorus through the pedon suggests that, minimally, direct sampling of buried horizons is required.

This methodology is most useful for regional studies that seek to determine the relative and overall (cumulative) importance of distinct landforms or ecological zones for human settlement. It is also useful for understanding spatial relationships between specific biophysical or political-economic parameters, on the one hand, and site properties and location, on the other. This spatial survey technique could also be employed to target site selection for more in-depth archeological and paleobotanical research, in which sites with distinctive properties or in different parts of the landscape could be sampled to deduce spatial and temporal patterns of past human activity. The utility of this integrative approach is demonstrated in the work of Mora et al. (1990) in the Caquetá region of Colombia.

Findings also shed light on the human ecology of blackwater regions by mapping the spatial distribution of Amerindian settlements in a region often described as restrictive to human settlement. Contrary to what is often expected, numerous anthropogenic soils dot these blackwater landscapes, suggesting that some form of semi-permanent settlement or frequent reoccupation characterized this region in the past. Expanding this study to other clearwater and whitewater regions would provide strong evidence for the relative importance of these highly distinctive environments in sustaining Amerindian populations in periods prior to and just following European arrival, a subject that has been strongly debated in recent decades (Meggers 1971; Roosevelt 1980; Heckenberger et al. 1999). The application of these methods to new geographical regions would require techniques to be refined so that the influence of distinctive parent material on soil coloration, phosphorus retention, or other indicators of anthropogenic impact be considered.

Applications of this methodology to the Lower Negro region shed light on important facets of human ecological theory. Deciphering the cumulative importance of any given geographical unit (landform, ecosystem, or region) for human settlement relative to other such units allows us to draw important associations between physiography, resource distributions, and human settlement. Contrary to the early literature stressing single limiting factors in blackwater regions (Meggers 1971; Gross 1975; Carneiro 1988), settlement patterns may be interpreted as maximizing access to heterogeneous or transitional environments where diverse resources may be exploited, and to strategic locations where political-economic control, defense, and transport are facilitated. These findings add complexity to materialist-environmental interpretations of human settlement distribution through the understanding of how diverse environmental, social, and political-economic factors may have together or independently conditioned resulting settlement patterns.

This recognition supports findings from the human ecological literature in which integrated economic strategies are the norm for contemporary and historical occupants (Posey 1985; Denevan and Padoch 1987; Anderson and Ioris 1992). The success of government-sponsored land reform programs could benefit a great deal if these understandings were incorporated into the design of settlement schemes. At present, these programs pay scant attention to the quality and heterogeneity of soils (Cravo, pers. comm.), and much less to the dependence of traditional inhabitants of Amazonia on the diverse and complementary resources that ensure the viability of their livelihood. The responsibility for bringing these findings to light within public policy circles, however, lies in the hands of the anthropologists, historians, and naturalists who have made Amazonia a professional home.

Acknowledgements. I would like to thank the *Secretaria de Cultura* of Amazonas State for funding. The *Fundação Vitória Amazônica* deserves special thanks for providing important logistical assistance during fieldwork and technical assistance during the analytical stage. I would also like to extend my thanks to Vivian Ziedemann for inspiration during the preliminary stage of the project, and to V. Zeidemann, Gerley Castro, Carlos Durigan, Marcos Pinheiro, and Jeff Walker for their help with fieldwork and technical assistance. Finally, I would like to thank Bruno Glaser and other participants of the 2001 Conference of Latin Americanist Geographers in Benicássim, Spain, who provided strong encouragement for the publication of findings.

References

Anderson AB, Ioris EM (1992) Valuing the rain forest: economic strategies by small-scale forest extractivists in the Amazon Estuary. Hum Ecol 20:337–369

Carneiro RL (1988) The circumscription theory. Am Behav Scientist 31:497–511

Chernela JM (1993) The Wanano Indians of the Brazilian Amazon. University of Texas Press, Austin

Denevan WM, Padoch C (1987) Swidden-fallow agroforestry in the Peruvian Amazon. Advances in economic botany, vol 5. New York Botanical Garden, New York

Eidt RC (1977) Detection and examination of anthrosols by phosphate analysis. Science 197:1327–1333

Ferreira Reis AC (1906) História do Amazonas. Editora Itatiaia Limitada, Manaus, Brazil

Goulding M, Leal Carvalho M, Ferreira EG (1988) Rio Negro, rich life in poor water. SPB Academic Publishing, The Hague

Gross D (1975) Protein capture and cultural development in the Amazon Basin. Am Anthropol 77:526–549

Heckenberger MJ, Petersen JB, Neves EG (1999) Village size and permanence in Amazonia: two archaeological examples from Brazil. Latin Am Antiq 10:353–376

Herrera R (1985) Nutrient cycling in Amazonian forests. In: Prance GT, Lovejoy TE (eds) Amazonia. Pergamon Press, Oxford, pp 95–105

Klinge H (1967) Podzol soils: a source of blackwater rivers in Amazonia. Liminologia 3:117–125

McCann JM, Woods WI, Meyer DW (2001) Organic matter and anthrosols in Amazonia: interpreting the Amerindian legacy. In: Rees RM, Ball BC, Campbell CD, Watson CA (eds) Sustainable management of soil organic matter. CAB International, Wallingford, pp 180–189

Meggers BJ (1971) Amazonia: man and culture in a counterfeit paradise. Aldine, Chicago

Mora CS, Herrera LF, Cavelier FI, Rodriguez C (1990) Cultivars, anthropic soils and stability: a preliminary report of archaeological research in Araracuara, Colombian Amazônia. Latin American Archaeology Reports 2, University of Pittsburgh, Pittsburgh

Moran EF (1991) Human adaptive strategies in Amazônian blackwater ecosystems. Am Anthropol 93:361–382

Nimuendaju C (1955) Reconhecimento dos Rios Içana, Ayari e Uaupés. J Soc Am 64:149–178

Pabst E (1991) Critérios de Distinção entre Terra Preta e Latossolo na Região de Belterra e os Seus Significados para a Discussão Pedogenética. Bol Mus Para Hist Nat Etnogr 7:5–19

Porro A (1987) O Antigo Comércio Indígena na Amazônia. Diario Leitura 5:2–3

Porro A (1992) História Indígena do Alto e Médio Amazonas: Séculos XVI a XVIII. In: Carneiro da Cunha M (ed) História dos Índios no Brasil. Editora Schwarcz, São Paulo, pp 175–196

Posey DA (1985) Indigenous management of tropical forest ecosystems: The case of the Kayapo Indians of the Brazilian Amazon. Agrofor Syst 3:139–158

Reichel-Dolmatoff G (1985) Tapir avoidance in the Colombian Northwest Amazon. In: Urton G (ed) Animal myths and metaphors in South America. University of Utah Press, Salt Lake City, pp 107–143

Ribeiro BG (1995) Os Índios das Águas Pretas. Companhia das Letras, São Paulo

Roosevelt AC (1980) Parmana: prehistoric maize and manioc subsistence along the Amazon and Orinoco. Academic Press, New York

Roosevelt AC (1992) Arqueologia Amazônica. In: Carneiro da Cunha M (ed) História dos Índios no Brasil. Companhia das Letras, São Paulo, pp 53–86

Salgado Vieira L (1988) Manual da Ciência do Solo, 2nd edn. Editora Agronômica Ceres, São Paulo

Salgado Vieira L, Oliveira Filho JPS (1962) As Caatingas do Rio Negro. Bol Tec Inst Agron Norte 42:1–32

Sioli H (1984) The Amazon: limnology and landscape ecology of a mighty tropical river basin. Junk, Dordrecht

Spaargaren OC (1994) World reference base for soil resources. Rome, Wageningen

Van Raij B (1991) Fertilidade do Solo e Adubação. Editora Agronômica Editora Agronômica Ceres, São Paulo

Woods WI (1975) The analysis of abandoned settlements by a new phosphate field test method. Chesopiean 13:1–45

Woods WI, McCann JM (1999) The anthropogenic origin and persistence of Amazonian Dark Earths. Yearbook Conf Latin Am Geogr 25:7–14

Wright R (1992) História Indígena do Noroeste da Amazônia: Hipóteses, Questões e Perspectivas. In: Carneiro da Cunha M (ed) História dos Índios no Brasil. Companhia das Letras, São Paulo, pp 253–266

5 Paleoriverine Features of the Amazon Lowlands: Human Use of the 'Arena Negra' Soils of Lake Charo, Northeastern Peru

OLIVER T. COOMES[1]

5.1
Introduction

Amazonia has long been divided by observers into two distinct environments – the upland (or *terra firme*) and the lowlands (or *várzea*) – a distinction that pervades the literature on subjects ranging from prehistory, archaeology, and Amerindian cultures to biological diversity, natural resource use, and agricultural development. The upland environment, lying above the floodplain of the Amazon River and its tributaries, offers a stable substrate, old forests, sparse game and soils of limited agricultural potential. In contrast, the lowland environment is highly dynamic and unstable, with young forests, abundant aquatic fauna, and nutrient-rich alluvium. Such depictions, however, are challenged by a growing body of research that points to high soil heterogeneity on the uplands (and some of high native fertility), the extensive occurrence of anthrosols – *terra preta* soils – on the Brazilian uplands (see chapters in this book; Lehmann et al. 2004), and the recognition of paleoriverine land forms that lie beyond the active floodplain of Amazonian rivers, but below the upland (Salo et al. 1986; Puhakka et al. 1992). Whereas *terra preta* soils appear to be frequently encountered along major river courses and in many interfluvial areas of the Brazilian Amazon, far fewer *terra preta* sites have been identified in the Upper Amazon and such sites tend to be found on river bluffs.

In this chapter, I suggest that paleoriverine features in the Amazon Basin are worthy of closer study, as potentially important sites of prehistoric agriculture and of significant potential for agricultural development. The paucity of *terra preta* sites in Upper Amazonia is perhaps not surprising given the proportionally large area of lowland and the dynamism of the Andean rivers that are continually reworking their floodplains (viz. Lathrap 1968, 1970). Nonetheless, extensive areas do exist in the lowland of paleo-floodplains. Salo et al. (1986) estimate that previous floodplains occupy about 75,100 km^2 or 14.6% of the area of Peruvian lowland forest. Abandoned floodplains – marked by paleomeanders, abandoned channels, scroll-bar complexes, islands, and levees – are found not only along the Ucayali, Marañón, and Amazon rivers, but also in association with tributaries such as the Pastaza,

---

[1] Department of Geography, McGill University, Montreal, H3A 2R6, Canada

Tigre, and Tapiche Rivers (see Puhakka et al. 1992; Pärssinen et al. 1996). Paleoriverine landforms would seem ideally suited for agricultural production – combining the higher fertility of lowland alluvial soils with relative security from frequent floods – and, if located near upland bluffs (à la Denevan 1996), are likely to have been extensively used by farmers in prehistory. For modern farmers, access to markets is also critical and paleoriverine features near rivers (or roads) that link producers with proximate urban centres offer high promise for market-oriented agriculture.

The case for closer attention to paleoriverine features in Amazonia is illustrated here through a study of a paleomeander island that lies within the Amazon River valley but above the meander belt floodplain and beyond the reach of decadal floods, near Lake Charo in northeastern Peru. Local residents refer to the site as the 'yarinales', after the presence of the ivory nut or yarina palm (Phytelephus macrocarpa). The soils of the yarinales – dark brown, fertile sandy loams – are distinct from those of both the Amazon River floodplain and the Tertiary upland. Since the late nineteenth century, farmers have practiced on the site one of the most productive and profitable forms of lowland swidden-fallow agroforestry yet described for the Amazon Basin. The following account of site characteristics and human uses of this paleo-island is based upon field information gathered between 1989 and 2000, during multiple visits to the area.[2]

5.2
Site Location and Characteristics

The paleomeander island is located at 4°15'S 73°15'W, near Lake Charo, about 65 km due south of the city of Iquitos in northeastern Peru (Fig. 5.1). At this point, the Amazon Valley spans a distance of some 20 km, with the Amazon River flowing along the western margin and its active floodplain dissected by recent scroll bars, inter-bar lakes, and streams. To the east, at slightly higher elevations, is a low floodplain terrace upon which the Tahuayo River flows, joining the Amazon River downstream, and we find the paleomeander island (see Räsänen et al. 1998). The island takes the form of a shield comprised of

---

[2] In 1989, I and my assistants conducted a detailed household survey in each of the 16 communities and two colonies in the basin, including the villages that use lands on the yarinales, to assess economic livelihood and market specialisation in agriculture and natural resource use (see Coomes 1992, 1998). The following year, we undertook field visits to the yarinales, collecting soil samples and assessing selected fields and orchards for cropping/fallowing history, crop choice and density and area. Records from boats transporting produce from the Tahuayo to Iquitos were gathered and used to assess the seasonality of production by villages along the river. In 1994, we returned to interview residents of the community of Esperanza – under which the primary jurisdiction of the paleomeander island falls – with respect to land holdings, avocado production, and non-agricultural uses; we visited the paleo-island once again with local authorities, discussed the history of land use, and dug test soil pits for description. In our last visit during 2000, we re-interviewed avocado producers in Esperanza and estimated production, assessed prices and costs, tree losses, and tree planting.

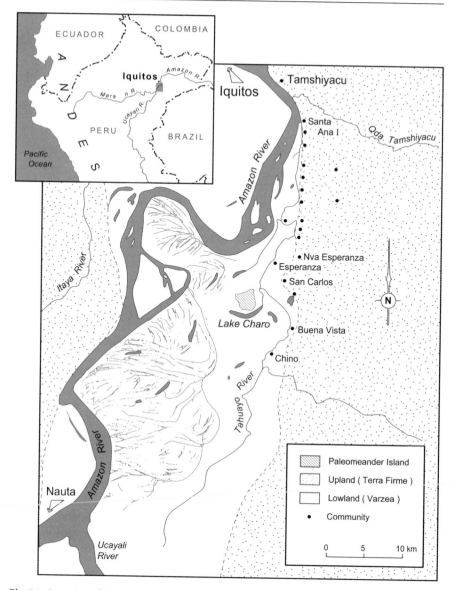

Fig. 5.1. Location of the paleomeander island, Lake Charo, Peruvian Amazon

low ridge-and-swale topography which covers a total area of about 635 ha and is raised above the surrounding lowland by about 2.5–4.0 m. Not subject to annual flooding, the highest lands of the site would seem to have been only rarely flooded over the past 100 years, if at all. The seral stage of tree species found on the site include *ojé* (*Ficus* spp.), *lupuna* (*Ceiba pentandra*), *huimba* (*Ceiba samauma*), and *quinilla* (*Pouteria* spp.). To the south, the oxbow lake

of Charo occupies an abandoned river channel of a meander that once
embraced the island. The genesis of the island is likely to resemble that
described by Mertes et al. (1996) for scroll-bar complex development in the
Brazilian Amazon.

The soils of the paleo-island complex are distinct from soils encountered
on either the upland or the lowland. The dark brown, sandy loams of the
yarinales – referred to by local farmers as *'arenas negras'* – are considered to
be particularly fertile. A hand-dug profile to 126 cm below the surface finds
the organic layer (0–23 cm) and brown sandy loam (23–58 cm); with increasing depth we observe reduced organic matter, decreased darkness, and
increased portion of mica-rich sand to a sandy basement. The physico-
chemical characteristics of the *yarinales* soils (0–15 cm) are distinct, as indicated by soil samples taken on the nearby upland (Ultisols) and along the
whitewater Amazon River floodplain (Fluvents) (see Table 5.1). The *yarinales*
soils are high in sand content and low in both silt and clay. The lack of flooding appears to have enabled soil formation and structural development.
Chemically, the pH is higher than the upland, but lower than soils of the
Amazon floodplain. Organic matter content is not distinct. Of the nutrients,
phosphorus stands out as being significantly higher – suggestive of past
human occupation and indicative of the higher fertility of the soils on this
'perched' island (Fig. 5.2). In areas inundated annually, a clay cap (about

Table 5.1. Mean physico-chemical characteristics of soil samples from agricultural fields
on the paleomeander island, Amazon River floodplain, and Tertiary upland, northeastern
Peru (0–15 cm). Samples were gathered in 1990 from fields on the *yarinales* (relic island),
from lowland fields in Tapira Nueva (Amazon River), and upland fields between Santa
Ana I and Punga, Tahuayo River (upland). In each field, an Oakfield sampler was used at
sites following an 'S' along the long axis of the field. Composite samples were derived,
bagged, air-dried in Iquitos, and transported to the Soils Laboratory at the University of
Wisconsin-Madison for analysis. Samples were extracted using water (for pH), Bray P1
(available P and K), and neutral ammonium acetate (exchangeable Ca and Mg)

Parameter	Meander Island	Amazon River floodplain	Tertiary upland
pH	5.7	7.3	4.0
Organic matter (%)	3.16	2.07	4.33
Phosphorus (ppm)	29.7	9.3	5.7
Potassium (ppm)	50	103	48
Calcium (ppm)	1,380	2,945	63
Magnesium (ppm)	177	266	30
Sand (%)	84	17	51
Silt (%)	13	66	34
Clay (%)	2	17	14
Number of samples	10	10	10

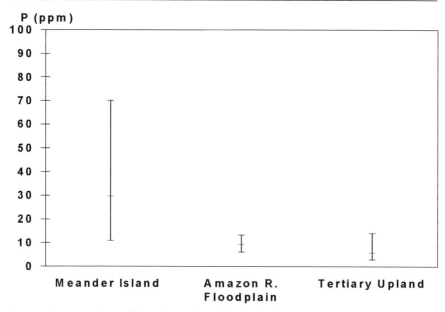

Fig. 5.2. Concentrations of phosphorus in soil samples (0–15 cm) from the paleomeander island, Amazon River floodplain, and Tertiary upland, northeastern Peru

50 cm thick) covers the land surface and soil properties are similar to that of the Tahuayo River floodplain (Gleysols).

Seven villages are located in the vicinity of Lake Charo, ranging in distance from 1–11 km to the paleomeander island. Historically, lands in the *yarinales* have belonged to residents of Esperanza, the largest and oldest of the communities in the region, about 7 km (or 1 h by canoe) downstream of the site on the Tahuayo River. Residents of Esperanza as well as the other villages have long cultivated agricultural fields on the *yarinales*, hunted game, extracted non-timber forest products, and collected medicinal plants from the forest. Lake Charo is also a site of an important local fishery, particularly as river levels fall in June and July.

5.3
Early Use of the *Yarinales*

Although not known to archaeologists as a site of prehistoric occupation, the location, site characteristics, and presence of material artefacts (potsherds and stone implements) all suggest the presence of humans during prehistory. Islands in the Amazon River are known to have been intensively cultivated in pre-history, especially between Iquitos and Pevas, but perhaps also upstream. As the meander was cut-off and abandoned, the secluded higher ground of the island near an oxbow lake would have been an attractive site – providing early residents with an abundant supply of fish, access to resources in the nearby, palm-rich swamp forest as well as to the secure upland where they could take refuge during exceptionally high floods. No reports are found in scholarly research on the location of Amerindian communities at contact in the region, nor is the area mentioned in early travel accounts despite its proximity to the mission of San Joaquin de Omaguas (founded in 1697) along the Amazon River. Some two dozen farmers with land on the *yarinales* reported to us their finding artefacts, including painted and unpainted potsherds and stone implements, though most recovered material is of historic rather than prehistoric origin.

In historic times, the forests of the paleomeander island have been worked since at least 1877 when the upper Tahuayo River Basin was a focus of wild rubber production (Coomes 1995). A land survey conducted in 1912 of the rubber estate of "Actividad" (4,854 ha) – which encompassed the *yarinales* – found 54 rubber trails (*estradas*), 200 ha of agricultural fields, and 60 ha of pasture.[3] Rubber tappers lived in isolated huts along the trails, tapping latex from rubber trees (*Hevea brasiliensis*) and practicing swidden-fallow agriculture for subsistence. The early tappers are said to have introduced avocado (*wira* variety) to the area. During the post-boom period (1920s–1940s) residents of nearby Esperanza harvested large quantities of vegetable ivory (up to 10 mt/month) for sale to a button factory in Iquitos. In the mid-1940s, when the demand for tagua fell with the advent of hard plastics and a brief revival of rubber ended with World War II, residents shifted their use increasingly to swidden-fallow agroforestry and began clearing the yarina palm for new fields. Such fields were opened only by the estate owners and their sons who introduced a larger variety of avocado (e.g., 20 cm, *palta grande*) and set aside a low-lying area of the *yarinales* as a reserve to conserve the supply of yarina palm fronds used for roof construction. Estate tenants were first allowed to hold fields in the *yarinales* in the 1960s, and in 1974 the estate was formally dissolved by the Agrarian Reform, leaving the land to the 'tillers'.

Today, the farmers who work the coveted lands of the *yarinales* are descendants of the original estate owners, former estate peones, and tenants, as well as newcomers. Of mixed Iberian and Amerindian origins, these mestizo people (*ribereños*) rely on a mix of traditional agriculture, fishing, hunting, and forest product extraction for their livelihood (see Hiraoka 1985; Padoch 1988; Chibnik 1994). Surplus and speciality crops are sent to market by daily river boats, a journey of about 1 day downstream to Iquitos, the primary urban centre and market in northeastern Peru. All are economically poor, earning typically less than US$ 600/year in cash income and holding 5–10 ha of land and less than US$ 300 of non-land assets (Coomes 1992). Although often described as highly egalitarian in terms of wealth (or pov-

---

[3] As reported in the *Informe Sobre la Demaración del Fundo* and the *Informe Pericia* both of November 1912, as part of the application by Sr. Rafael Pinedo Rios for title to the property "Actividad". Title granted on 2 November 1918 (no. 1068) and cancelled in 1974 by *Resolución Directoral Zonal* no. 163-DZA-VIII-74.

erty), recent studies suggest that land and other economic assets can be highly unequally distributed within traditional riverine communities (Coomes and Burt 1997; Takasaki et al. 2001). In 1994, a total of 63 households from 6 neighbouring communities held land on the former island with a total of some 200 fields, encompassing a cultivated area of 100–150 ha.

5.4
Contemporary Agriculture

Farmers practice three swidden-fallow cycles on this paleo-landform – defined by the dominant crop and relative elevation – (1) a manioc/maize cycle; (2) a plantain cycle; and (3) an avocado orchard cycle (Fig. 5.3). In low-lying areas, where the depth of the annual flood is typically greater than 75 cm, farmers practice short cycle cultivation of manioc and/or maize. As floodwaters recede in May and June, farmers plant manioc and maize in these clayey soils and harvest before the flood returns in late December, leaving then the field in fallow for 2–3 years. Farmers will intercrop the maize and manioc with watermelon, yams, dale dale, peach tomato, vegetables (e.g., sweet pepper, tomatoes), and papaya. In selected places, rice is mono-cropped. At higher elevations on the *yarinales*, farmers cultivate manioc/maize, followed by plantain (3–6 years), and a fallow of 3–4 years. Plantain can withstand only temporary inundation of its stem and so requires somewhat higher land (i.e., <75 cm of floodwater). Interspersed among the plantain are yams, taro, bijao, papaya, and balsa wood. Yields of plantain here are reportedly double those on the upland. At elevations above floodwaters, where the soils are loamy, friable, and more fertile, farmers incorporate avocado, after manioc/maize and plantain, as the terminal tree crop, managing this cycle over a period of 25–60 years. Avocado orchard size ranges from 0.12–3.0 ha (median: 0.5 ha) with tree densities of 5–600 ind./ha (median: 70) and ages of 3–50 years (median: 25 years) (n=35 orchards). Alternate terminal tree crops include cacao, mango, sour sop, sapote, star apple, breadfruit,

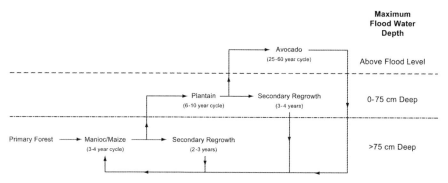

Fig. 5.3. Swidden–fallow agricultural cycles practiced on the paleomeander island, Lake Charo, Peru

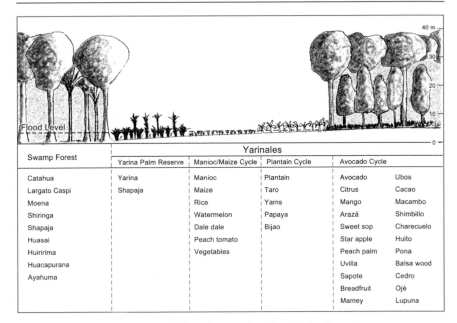

Swamp Forest	Yarinales				
	Yarina Palm Reserve	Manioc/Maize Cycle	Plantain Cycle	Avocado Cycle	
Catahua	Yarina	Manioc	Plantain	Avocado	Ubos
Largato Caspi	Shapaja	Maize	Taro	Citrus	Cacao
Moena		Rice	Yams	Mango	Macambo
Shiringa		Watermelon	Papaya	Arazá	Shimbillo
Shapaja		Dale dale	Bijao	Sweet sop	Charecuelo
Huasai		Peach tomato		Star apple	Huito
Huririma		Vegetables		Peach palm	Pona
Huacapurana				Uvilla	Balsa wood
Ayahuma				Sapote	Cedro
				Breadfruit	Ojé
				Mamey	Lupuna

Fig. 5.4. Schematic cross section of the paleomeander island, Lake Charo, Peru

macambo, arazá, and even citrus and umarí; indeed, the only crops known not to grow well here are pineapple and cashew, both crops that prefer acidic soils. Only the fields of maize/manioc and of banana are visible from the air because the avocado orchards are surrounded by taller native trees that afford the orchards protection from windthrow (see Fig. 5.4).

Avocado is the 'signature' crop of the *yarinales*, by far the most important cash crop to local farmers and well known to buyers in the markets of Iquitos. Ideally suited in the swidden-fallow agroforestry cycle, avocado fetches a good and relatively stable price (US\$ 10–15/100) and is rich in protein, serving as an important dietary supplement, especially among children. The tree has a long productive life and coppices easily – if the tree is blown over, farmers will cut the trunk and re-seat the base (which coppices forth a new tree) or leave the fallen bole which coppices from the ground, an important advantage over other potential tree crops. The fruit is large and requires ripening, typically at home, before shipping to market; two features that reduce large-scale theft from these distant fields. The tree, however, is highly sensitive to water, perishing if flood waters reach its roots and vulnerable by its shallow root system in these sandy soils to windthrow. Paca prey on the seeds and a tree borer (*gusano*) cripples mature avocado.

Two varieties of avocado are produced – the smaller and finer, *wira palta* and the thicker-skinned, *palta grande*. The tree is raised from seedlings, collected from beneath mature avocado, and transplanted at distances of at least 5 m from one another so as to limit the need for the tree to grow tall. Undergrowth is left around the base of the tree in order to preserve soil moisture. Avocado here bears fruit first in the sixth year after planting and has a productive life of about 35–40 years, although we observed two trees that were reputedly 80 years old. Each tree will produce between 200 and 500 fruit/year (*palta grande*) and 400–700 fruit/year (*wira palta*) during two seasons – mid-April to the end of July and October to December. Although only a fraction of production is sold (i.e., the highest quality fruit), we estimate market production from the area during the 1990s to have been 75,000–200,000 fruit/year from 40 producers, of whom 25 reside in Esperanza and harvest at least 80 % of the total crop.

Avocado is both an important source of cash income for farmers and a vital economic asset. Our survey among Esperanza producers in 2000 found sales for the previous year of 350–26,000 fruit/household (mean: 5,100) for an estimated gross value of about US$ 15,000 (Table 5.2). Although the costs of harvesting and transportation are significant (about 30 % of gross), households are left with substantial income, particularly when the incomes of large producers (US$ 1,000–3,000/year) are compared with the rural wage of US$ 1–2 /day. Indeed, earlier survey data (1988–1989) from the same producers indicated that avocado sales contributed a significant amount of cash income to producing households in the community and both their incomes and total land holdings were significantly higher than non-producers. Production though is quite highly concentrated, with the top five producers harvesting just over one-half of the total crop (53 %). The top producers are descendants of the community founders and whose families have lived well for as many as five generations from avocado, sending their children to Iquitos for schooling and, in good years, purchasing much of their food. Although *yarinales* land per se is held only in usufruct (i.e., it cannot be bought or sold), individual avocado trees are sometimes purchased (US$ 10–15/tree) and households consider them as liquidable assets that can be passed on to their children. Again, better-off households have substantial orchards, ranging up to 75–115 trees, with larger holdings on the highest land.

Table 5.2. Avocado production in Esperanza, Tahuayo River, Peru, 1999–2000. Source: interviews with avocado producers ($n=25$)

	Palta Grande	Wira Palta	Total
Estimated production (no. of avocados)	84,750	41,250	126,000
No. of avocado trees (no. of households)	859 (25)	223 (19)	1,082
No. of trees lost, 1997–2000	786	297	1083
Percentage lost to flood/windthrow	80	63	
Percentage lost to tree borer	20	37	
No. of trees coming into production, 2000–2003	737	475	1,212

As such, and interestingly, the vertical stratification of land use on the *yarinales* – into distinct swidden–fallow cycles by elevation – corresponds directly with socio-economic stratification among households in Esperanza. The founding families hold more land in the *yarinales*, higher land, and larger orchards; later arrivals hold less land (or none in the *yarinales*), at lower elevations, and smaller (or no) orchards. Over time, the founding families consolidated their early claims, persisting with profitable avocado (which financed deeper investment in orchards as well as in higher education), whereas later arrivals (with lower *yarinales* lands) were more vulnerable to periodic high floods – ultimately the latter relinquished their land or shifted to the less profitable swidden–fallow cycles around maize/manioc and banana. In this manner, the lands of the paleomeander island have been instrumental in the evolution of wealth differentiation among peasant households in Esperanza since the *yarinales* were first claimed and worked in the late nineteenth century.

5.5
Non-Agricultural Resource Use

Though of lesser importance than to swidden–fallow agriculture, the *yarinales* are a significant source of non-timber forest products, game, and medicinals. Building materials, especially fronds of the yarina palm for hut roofing, are drawn from the area, especially the yarina reserve which is protected by households from Esperanza though only lightly managed (i.e., clearing around base when culling fronds). Game including deer, paca, agouti, peccary, monkeys, and a variety of bird species are hunted on the *yarinales*, some drawn by the food available in the fields and orchards. During high floods, game tends to concentrate on the high ground and is particularly vulnerable to hunting. Medicinals collected from the forests of the *yarinales* include *lancetilla* (shrub), *ishanga* (tree), *citualla* (cane), *cituje* (tree), *shucahuito* (tree), and *requia* (tree).

5.6
The Challenges of Future Use

Four challenges beset farmers who work the lands of this paleo-landform. First, the avocado orchards need to be continually renewed. In 2000, producers in Esperanza reported 1,082 avocado trees in production in the *yarinales*, and during the previous 3 years they had lost an equivalent number to flooding/windthrow (75%) and pest damage (25%) (Table 5.2). Producers projected that 1,212 trees would come into production between 2000 and 2003, or about 12% more than lost. Planting tends to occur in spurts, typically after 2–3 years of low annual floods, and ceases after a particularly high flood when losses are greatest. Clearly, orchards on somewhat lower land are being renewed, by force of tree mortality, but orchards on the highest grounds –

where many trees are of similar age, often 20–30 years or more old – are not. Standing orchards are renewed with difficulty because young avocado trees require abundant light and seedling mortality due to agouti is high. Typically, residents must wait until an old, poor-producing tree falls and opens a gap before replanting occurs.

Second, farmers are facing increased competition in Iquitos markets from producers elsewhere. Prior to the early 1980s, avocado produced in the Maniti River Basin (down-river from Iquitos) competed with those from Esperanza and Lake Charo, but a large flood in 1982 appears to have significantly reduced orchard production in the region. Today, competition comes from two sources – communities such as Shebon and Perlita, along the Ucayali River, where high levees and *'yarinales'* are found close to the channel, and from upland communities along the Iquitos–Nauta road and Itaya River. Avocados from these sources are reputed to be smaller and less refined, selling for 50 % of the price of Tahuayo avocado. In general, avocado production on the *terra firme* is short lived (1–3 years of fruit production) and requires significant inputs to offset soil poverty, moisture deficits, and the ravages of leaf cutter ants.

A third set of challenges arises in conflicts over land on the *yarinales*, particularly between the communities of Charo and Esperanza. Charo was established in the 1970s by people displaced by a high flood along the Amazon River, as a base for extraction of monkeys from the *yarinales*, fish from Lake Charo, and moriche palm fruit from the swamp forests to the south. Although Charo is much nearer to the *yarinales* than Esperanza, claims over the lands on the paleo-island have been held by families of Esperanza for generations. With little agricultural land available around the village site and having depleted the local stock of monkeys and moriche palm fruit, residents of Charo sought land in the *yarinales*, with only limited success. In December 1991, however, (ex) President Fujimori accompanied by five members of Congress paid an unexpected visit to the lake (a popular sports fishing destination) by float plane. Much impressed by Charo residents' self-representation as 'conservationists' of the nearby *yarinales*, the ex-president unwittingly decreed that the community should be granted title over the *yarinales*. This decree triggered an intense land struggle, but was ultimately disallowed by the regional government, though conflicts persist. In the late 1990s, residents of the community of San Carlos (well inland from the Tahuayo River) sold their orchards to residents of Charo, reportedly because they could no longer afford the effort to prevent theft.

Finally – and perhaps most seriously – the flood regime in the region appears to be changing as the Amazon River began to migrate laterally (eastward) in the 1980s and threatens to re-take the low terrace (see Kalliola et al. 1992; Tuukki et al. 1996). Residents of Esperanza reported in 1982 that the high levee that separated the Amazon and Tahuayo River systems near Yacapana was breached by the Amazon River, and flood waters from the Amazon flowed through Lake Charo and into the Tahuayo River. In 1989, downstream on the Tahuayo River at Huaisi, the Amazon River broke through another levee and captured the lower Tahuayo along a reach of 13 km. Floodwaters from the Amazon River now cut across Lake Charo for up to 3 months of the year and year-round below Huaisi. High floods in 1993 and 1994 took a heavy toll on the avocado orchards, with production falling from 157,000 in 1988–1989 to only 75,500 in 1993–1994; by 1999–2000, however, avocado output had recovered to production levels of the late 1980s. In time, residents are concerned that the changing course of the Amazon River may not only bring a new dominant flood regime (i.e., that of the Amazon rather than the Tahuayo River), but also possibly destroy the *yarinales* and their profitable agroforestry system.

5.7
Conclusions

In this chapter I have described a paleoriverine feature in the Amazon River lowland (*várzea*) of northeastern Peru where farmers practice a highly productive, sustained, and profitable form of lowland swidden-fallow agroforestry. The site – a paleomeander island covering an area of some 635 ha on a low terrace above the meander belt floodplain of the Amazon, with its *'arena negra'* soils – constitutes an ideal site for agricultural production, combining the high fertility of lowland environments with security from decadal flooding . Avocado – a nutrient-demanding and highly water-sensitive crop species – has thrived on the paleo-island as the terminal tree crop for almost a century and provided significant income for peasant farmers for several generations. Elevated phosphorus concentrations in the soil and the presence of potsherds and stone implements suggest prehistoric use and occupation of the site. Are these *'arena negra'* soils though *terra preta*? Perhaps not; the paleo-island is clearly a depositional feature and the brown, sandy loams are the likely product of long-term additions of organic matter from the ivory nut palm and phosphorus-rich human detritus to the young Andean alluvium that forms the core of the island. Nevertheless, the high fertility of the soils and the promise for sustainable agricultural development beg further study. The common occurrence of paleoriverine features in the heterogeneous lowlands of the Peruvian Amazon – including paleomeander islands, terraces, high levee fragments, and perched scroll bars – and reports of avocado production elsewhere in the lowlands suggest that similar sites of high agricultural potential and past human use are to be found in the Upper Amazon.

Acknowledgements. The author gratefully acknowledges the farmers of Esperanza for so willingly participating in this study, particularly Don Carlos Rivas, Cesar Rivas, and Francisco Huayllahua. This research was supported by grants from the Social Sciences and Humanities Research Council of Canada, Inter-American Foundation, University of Wisconsin-Madison, the World Wildlife Fund, and the *Fonds pour la Formation de Chercheurs et l'Aide à la Recherche*. Invaluable field assistance was provided by Jomber Chota Inuma and Carlos Rengifo Upiachihua. Antoinette WinklerPrins kindly facilitated the analysis of soil samples at the UW-Madison Soils Laboratory.

References

Chibnik M (1994) Risky rivers: the economics and politics of floodplain farming in Amazonia. University of Arizona Press, Tucson

Coomes OT (1992) Making a living in the Amazon rain forest: peasants, land, and economy in the Tahuayo River basin of northeastern Peru. PhD Diss, University of Wisconsin-Madison, Madison, Wisconsin

Coomes OT (1995) A century of rain forest use in western Amazonia: lessons for extraction-based conservation of tropical forest resources. For Conserv Hist 39(3):108–120

Coomes OT (1998) Traditional peasant agriculture along a blackwater river of the Peruvian Amazon. Rev Geogr 124:33–55

Coomes OT, Burt GJ (1997) Indigenous market-oriented agroforestry: dissecting local diversity in western Amazonia. Agrofor Syst 37(1):27–44

Denevan WM (1996) A bluff model of riverine settlement in prehistoric Amazonia. Ann Assoc Am Geogr 86(4):654–681

Hiraoka M (1985) Mestizo subsistence in riparian Amazonia. Natl Geogr Res 1(2):236–246

Kalliola R, Salo J, Puhakka M, Rajasilta M, Häme T, Neller RJ, Räsänen ME, Danjoy Arias WA (1992) Implications for vegetation perturbance and succession using bitemporal Landsat MSS images. Naturwissenschaften 79:75–79

Lathrap DW (1968) Aboriginal occupation and changes in river channel on the central Ucayali, Peru. Am Antiq 33:62–79

Lathrap DW (1970) The Upper Amazon. Praeger, New York

Lehmann J, Kern D, Glaser B, Woods W (2004) Amazonian Dark Earths: origin, properties, management. Kluwer, Dordrecht

Mertes LAK, Dunne T, Martinelli LA (1996) Channel–floodplain geomorphology along the Solimões-Amazon River, Brazil. Geol Soc Am Bull 108(9):1089–1107

Padoch C (1988) People of the floodplain and forest. In: Denslow JS, Padoch C (eds) People of the tropical rain forest. University of California Press, Berkeley, pp 127–141

Pärssinen M, Salo JS, Räsänen ME (1996) River floodplain relocations and the abandonment of aborigine settlements in the Upper Amazon basin: a historical case study of San Miguel de Cunibos at the middle Ucayali River. Geoarchaeology 11(4):345–359

Puhakka M, Kalliola R, Rajasilta M, Salo J (1992) River types, site evolution and successional vegetation patterns in Peruvian Amazonia. J Biogeogr 19:651–665

Räsänen M, Linna A, Irion G, Rebata Hernani L, Vargas Huaman R, Wesselingh F (1998) Geología y geoformas de la zona de Iquitos. In: Geoecología y desarrollo Amazónico: estudio integrado en la zona de Iquitos, Perú. Ann Univ Turk Ser A II 114:59–137

Salo JS, Kalliola R, Hakkinen I, Makinen Y, Niemela P, Puhakka M, Coley PD (1986) River dynamics and the diversity of the Amazon lowland forest. Nature 322:254–258

Takasaki Y, Barham BL, Coomes OT (2001) Amazonia peasants, rain forest use, and income generation: the role of wealth and geographical factors. Soc Nat Resour 14:291–308

Tuukki E, Jokinen P, Kalliola R (1996) Migraciones en el río Amazonas en las últimas décadas, sector confluencia ríos Ucayali y Marañón – Isla Iquitos. Folia Amaz 8(1):111–131

6

Dark Earth in the Upper Amazon

Thomas P. Myers[1]

6.1
Introduction

The dominant culture historical model of Amazonian prehistory is that population growth in the central Amazon led to repeated migrations up the major tributaries including the Madeira, Japurá, Napo, and Ucayali Rivers. Lathrap argued that this population growth was made possible by the extraordinarily rich *várzea* of the central Amazon. In time, overpopulation, internecine warfare, and/or other factors led some to emigrate in search of comparable lands elsewhere (Lathrap 1970; Brochado 1984).

Accumulating research demonstrates that many, if not most or all, important prehistoric occupations on the central Amazon are distinguished by dark, highly fertile soils that are consistently associated with large concentrations of ceramic, lithic, faunal, and botanical remains. Charcoal is usually abundant. These are known as *terra preta* (TP) or black earth. These black earths may have been a critical component of the central Amazonian adaptation. TP is highly productive. Once created, it retains a high productivity for hundreds, even thousands of years. Furthermore, it is self-regenerating. Yet, we do not know how it is formed. Was it a by-product of human occupation or was it the result of specific, goal-directed, activities of the prehistoric inhabitants?

This chapter begins by examining the living and sanitation circumstances in modern Amazonian villages to ascertain whether black earth formation is likely to be an accidental by-product of village life. It also tests the hypothesis that dark earth technology was the critical innovation that facilitated a population explosion in the central Amazon by examining diverse regions to determine whether local residents produced dark earth prior to the immigration of peoples from the central Amazon.

---

[1] University of Nebraska State Museum, W512 Nebraska Hall, Lincoln, Nebraska 68588-0514, USA

6.2
The Formation of Dark Earth Sites

Nearly everyone agrees that dark earth sites are the product of human activity, but there is some question about just how they are formed. Implicit or explicit, there are two underlying hypotheses. Many believe that dark earth sites are the accidental by-products of human habitation. Some are explicit that this is the case. Others believe that dark earth was produced by a specific technology, or sets of techniques, intended to produce permanently productive soils on the *terra firme* in a tropical forest environment (Herrera et al. 1992).

For one who has lived in several Amazon Indian villages, it is difficult to understand how tens of centimeters of detritus could accumulate in occupied areas. The habitation zone, consisting of houses and streets or plazas, is typically surrounded by brush and second growth forest as well as by gardens and hunting grounds. House, street, and plaza areas are kept relatively clean. Parts of them are swept daily. The surface is being worn away in these areas rather than built up by the accumulation of new refuse. Prehistoric artifacts that have been buried for hundreds or thousands of years reappear on the surface.

Refuse from the streets, plazas, and houses is normally tossed into the brush surrounding the settlement. Such a community pattern would produce a ring-shaped site with little refuse in the center, a ring of refuse featuring broken pottery, and then an outer ring with sparse signs of human activity (Lathrap 1962; Myers 1973). In a very large site, houses might be interspersed with kitchen gardens, leaving patches of regularly cleaned areas interspersed with patches of household refuse where the kitchen gardens had been. Over the life of a village the patchwork might change if people were willing to destroy their kitchen gardens, which seems unlikely.

Consider San Francisco de Yarinacocha, a village occupied by several hundred Shipibo at least since the 1920s. The locality has been intermittently occupied for the last 4,000 years. A photograph of the village taken in 1987 (Fig. 6.1) reveals a main street virtually clear of litter, though closer inspection reveals prehistoric materials eroding out of the ground. Thatched houses with dirt floors line either side of the street except where there is a grassy space between the houses. These grassy spaces are the locale of past and/or future houses. Grass grows to the edge of a couple of abandoned houses on the right, but otherwise the cleared area extends from the street through the houses, and for some distance behind. Certainly, the main street does not seem a likely candidate for the formation of *terra preta* since the area is regularly cleaned to the point of being deflated. *Terra preta* might form in the grassy areas between the houses, but the grass and brushy areas behind the houses would seem to be prime candidates for *terra preta* formation. In 1956, Lathrap noted that as a general rule there is no accumulation of Shipibo midden within 8 m of a Shipibo house (Lathrap 1962).

Fig. 6.1. Main street of San Francisco de Yarinacocha in 1987

Some parts of San Francisco are not so scrupulously cleaned (Fig. 6.2). This isolated house is set back in the woods rather than being part of the main village. The raised floor suggests that it may be subject to flooding. Litter is found throughout the yard. Clumps of grass and algae and all sorts of litter come right up to the house. Cooking pots, pans, and buckets are scattered about. The abandoned fire in the background may have been for firing pottery since the photograph was taken in October, during the last days of the dry season. Substitute ceramics for the plastic and metal containers seen in this photo, and sherds would litter the ground. All the components for the formation of *terra preta* are present.

A Bora house in the community of Brillo Nuevo on the floodplain of the Río Ampiyacu (Fig. 6.3) is also surrounded by low grass that probably conceals a small quantity of litter, but not the volume seen near the isolated house in San Francisco. The raised floor of the house suggests that floodwaters are expected. While clear of brush and vegetation, the area beneath the house is subject to refuse accumulation. If the deep floodwaters, suggested by the height of the floor above ground, did not run too rapidly there might be a fairly rapid midden accumulation with cultural remains rather widely dispersed.

At Achual Tipischa (Fig. 6.4), a Cocamilla village on the floodplain of the Huallaga River, houses are again raised above the levels of the floodwater. The area immediately around the houses is cleared, sometimes with as much as 10 m of cleared ground in front of the houses. However, the areas between

Fig. 6.2. Shipibo house somewhat removed from the main street

Fig. 6.3. Bora house on Ampiyacu River, 1987

Fig. 6.4. Achual Tipischa, a Cocamilla village on the Huallaga River, 1980

the houses are grass covered. The ground beneath some of the houses is slightly raised due to deflation around them. Cultural refuse is visible in both cleared areas and grassy areas. Another view (Fig. 6.5) reveals erosion is also part of site formation. In this case, a wooden mortar has rolled into the depression. Because of the long vertical crack, it will soon be discarded. Other litter is visible nearby as is the 'house platform' created by soil deflation. The image suggests that there may be a slight darkening of the soil in front of the house.

Historic communities of the northwest Amazon usually consist of a single multi-family dwelling or *maloca* which may be 50 m in length and may house as many as 200 individuals (Domville-Fife 1924). A Kaua *maloca* photographed near the Aiary River in 1910 stands in a clearing about 40 m in diameter. In front of the house is a cleared plaza that extends outward for a distance of 10 or 15 m. Beyond that is short brush that comes right up to the house on both sides. The women's plaza at the back of the house may be less carefully cleaned than the front plaza, which may have been specially cleaned for the ceremonial occasion. Bará and Tuyúka *malocas* on the Tiquié River are similar (Koch-Grünberg 1910). According to Goldman, well-run *maloca* communities of this sort are clean and orderly. Houses are swept regularly and maintained in good repair. The plaza, the canoe landing, and the path between the canoe landing and the house are clean and in good order. The people defecate in the brush that lies some distance from the house. Dark earth might form in accumulation of kitchen debris at the rear of the house.

Fig. 6.5. Erosion in Achual Tipischa

A trash pile near the riverbank receives household debris and the body wastes of very young children, but the accumulated rubbish is washed away during the rainy season, so dark earth would have little chance to develop (Goldman 1963).

Jívaro communities of the twentieth century normally consist of a single dwelling that may shelter 10 to 30 or 40 individuals, rarely more. Communities of several houses may have been more common in the past. Oval houses may be 20–30 m long and 15–25 m wide (Thomas 1922; Domville-Fife 1924; Castrucci 1905). An isolated Jívaro house built on a hilltop offers rather different conditions for *terra preta* formation (Blomberg 1952). A cooking area behind the house is semi-cleared for a distance of perhaps 3 m, but at the sides of the house grass and weeds come within about 50 cm of the wall. A substantial amount of charcoal might be expected from the outdoor cooking fire as well as from the fires built within the house. Periodically, the ashes would be discarded into the brush behind the back door. Broken pottery and other household refuse would be similarly discarded. Once abandoned, the roof and walls of the house would collapse on top of the remains of indoor fires and the accumulated refuse beneath the sleeping benches.

South of the Amazon, communities commonly are formed of a circle of houses around a central plaza. Some are very large and may have been much larger in the past (Heckenberger 1996). An air photograph of a Waurá village on the upper Xingu reveals a cleared central plaza where there is unlikely to be any black earth formation. In contrast, the area behind the house might catch a considerable amount of litter from the production of manioc flower and other household activities. Collapsed houses would again yield a concentration of litter. Seen from the air, a Kraho village on the Tocantins River, some distance to the east, resembles a giant wheel with a cleared hub, spokes, and rim. The houses would collapse into litter, perhaps a little denser than the brushy areas between and behind the houses.

In whatever part of the Amazon we look, there is a minimal accumulation of refuse in the central portion of habitation sites. Only at the edges of the site is there a buildup of litter from which dark earth might be produced. Unless there were a substantial accumulation of litter over an extended period of time it is difficult to understand how dark earth might be produced without intentional composting activities. Even then, the black earth would be found at the edges of the site rather than as a thick deposit throughout the village. Black earth would only be found on top of a habitation site if the site had been abandoned prior to the spreading of the compost. This suggests that dark earth sites were intentionally formed as gardens rather than as an accidental by-product of human habitation. Andrade reached the same conclusion on the basis of soil chemistry (Andrade 1986).

6.3
Dark Earth Sites and Settlement Density on the Central Amazon

Dark earth sites covering as much as 80 ha appeared on the central Amazon by A.D. 200 when they were associated with the Barrancoid or Incised Rim tradition. Such sites continued to be formed and occupied for the next 1,300 years, through the hegemony of the Polychrome and Incised and Punctate traditions, into the time of European contact. The size and depth of the occupation layers leave little doubt that if these were habitation sites, many of them were occupied by several thousands of people for extended periods of time. Alternatively, they may have been gardens which supplemented the crops grown on islands in the *várzea* (Myers 2004; Myers et al. 2004).

It is uncertain how many of these sites were occupied simultaneously, or how many sites remain to be discovered, even in parts of the central and lower Amazon that are fairly well known. Consequently, we do not really know how dense the population actually was at any point in time. We do know, however, that villages or towns were densely packed along the high ground on the southern side of the Solimões between the Juruá and Negro Rivers at the time of European contact in the mid-1500s. The Machifaro town of Moçomoco, located in the midst of great savannas on the southern shore of the Amazon near Tefé, was large enough that the thousand members of the Orsua expedition could be billeted in one end of the settlement (Hernandez 1981; Ortiguera 1981; Vázquez 1981; Zuñiga 1981; Myers 2004). Towns closer to the Purús River were said to extend for a distances of 2–2.5 leagues, about 8.5–10 km (Myers 2003; Rowlett 2003).

In contrast, few recorded dark earth sites are more than a kilometer in length. One of the largest is the Manacapuru site (AM-MP 1), which stretches for a distance of 2,000 m along the bluffs on the northern shore between the Purús and Negro Rivers (Hilbert 1968; Simões and Araujo-Costa 1978). The site is certainly not far from the large towns of the sixteenth century. The size differential between the archaeological sites and the historic towns may mean that only some portion of the towns became dark earth.

Until recently, we had supposed that the very large aboriginal towns of the central Amazon had been supported primarily by cultivations on the seasonally flooded lowlands, or várzea, including the islands in the river. Certainly such várzea localities were extensively cultivated in the sixteenth and seventeenth centuries, just as they are today. In 1651, the gardens of the Aysuari, who lived between the Juruá River and Lake Tefé, were located on the islands even though the principal towns and villages were located on the terra firme of the southern shore (Vázquez 1981; Cruz 1999). We had always known, or at least supposed, that these seasonal island gardens were supplemented by gardens on terra firme. We had not realized that gardens placed on black earth sites or terra preta were essentially permanent. Nor had we realized that a good imitation of terra preta could be produced by the addition of human or animal manure, green manure from pounded mollusk shells, and/or root accretions (Woods and McCann 1999; Kern et al. 2004).

6.4
Barrancoid Migrations Out of the Central Amazon

While the evidence is far from complete, the facts that we can bring to bear strongly suggest that the central Amazon was densely occupied by Barrancoid peoples on the central Amazon by 200 B.C. and remained dominant in the region until A.D. 1050 (Table 6.1). During this period, competition for suitable lands was likely to have been growing ever more intense. Increasingly complex political organizations were almost certainly developing in the context of growing community size and population density (Carneiro 1970; Forge 1972; Myers 1992).

According to the Lathrap model, population growth and increased competition for scarce lands led some Barrancoid populations to leave the central Amazon, moving up the principal tributaries including the Madeira, Japurá, and Ucayali Rivers (Fig. 6.6). They seem to have reached the central Ucayali, where they are identified with the Hupa-iya Complex, perhaps as early as 300 B.C., even before the first radiocarbon dates from the central Amazon itself (Lathrap 1970; Brochado 1984). Barrancoid peoples first appear on the Japurá

Table 6.1. Calibrated (1δ) radiocarbon dates of the Barrancoid tradition. Calibrated using CALIB v4.3

Site	Location	Depth (cm)	Date (years B.P.)	1σ	Probability	Lab no.
Curralinho	Lower Madeira		500±55	A.D. 1331–1340	0.087	
				1397–1449	0.913	
Hatahara	Central Amazon	155	960±40	A.D. 1022–1060	0.397	Beta
				1086–1122	0.391	143595
				1138–1156	0.213	
Hatahara	Central Amazon	180–190	1,080±40	A.D. 898–921	0.305	Beta
				945–946	0.019	143598
				956–1002	0.637	
Hatahara	Central Amazon	160–170	1,070±70	A.D. 893–1022	1.000	Beta 143596
Santa Sofía	Trapecio de Amazonas		1,040±90	A.D. 891–1044	0.814	I5775
				1089–1121	0.119	
				1139–1156	0.067	
Jatapu	Lower Amazon		1,030±75	A.D. 897–922	0.136	
				942–1043	0.672	
				1092–1119	0.121	
				1140–1154	0.072	
Curralinho	Lower Madeira		1,065±90	A.D. 784–788	0.011	
				832–837	0.016	
				876–1040	0.893	
				1098–1116	0.049	
				1141–1152	0.030	
Curralinho	Lower Madeira		1,110±60	A.D. 783–789	0.033	
				829–839	0.057	
				864–986	0.910	
Mangueiras AM-JA-3	Japurá/Caquetá	Cut 1, 60–75 cm	1,315±59	A.D. 622–628	0.006	P588
				638–784	0.873	
				788–832	0.069	
				837–876	0.052	
Oswaldo	Central Amazon	41	1,290±30	A.D. 685–722	0.556	Beta
				742–770	0.444	143609
Oswaldo	Central Amazon	45	1,290±40	A.D. 680–726	0.559	Beta 143613
				738–773	0.441	Beta 143613
Hatahara	Central Amazon	192	1,300±40	A.D. 671–721	0.651	Beta
				743–769	0.349	143599
Oswaldo	Central Amazon	76	1,310±40	A.D. 664–694	0.426	Beta
				696–718	0.298	143620
				747–767	0.276	
Oswaldo	Central Amazon	36	1,330±40	A.D. 658–693	0.619	Beta
				699–715	0.202	143612
				750–763	0.179	
Oswaldo	Central Amazon	73	1,320±60	A.D. 657–725	0.687	Beta
				739–772	0.313	143619

Table 6.1. (Continued)

Site	Location	Depth (cm)	Date (years B.P.)	1σ	Probability	Lab no.
Oswaldo	Central Amazon	61	1,340±40	A.D. 654–692	0.727	Beta
				700–713	0.147	143617
				751–762	0.126	
Oswaldo	Central Amazon	50–60	1,350±30	A.D. 654–688	1.000	Beta
						143616
Oswaldo	Central Amazon	66	1,350±40	A.D. 645–691	0.877	Beta
				703–709	0.061	143618
				753–758	0.061	
Oswaldo	Central Amazon	42	1,360±50	A.D. 624–626	0.014	Beta
				639–692	0.802	143614
				700–713	0.096	
				751–761	0.088	
Oswaldo	Central Amazon	35	1,370±40	A.D. 624–625	0.010	Beta
				639–688	0.990	143610
Pedrera	Japurá/Caquetá		1,390±75	A.D. 564–571	0.036	SI6375
				578–588	0.049	
				597–692	0.815	
				701–712	0.057	
				751–761	0.043	
Oswaldo	Central Amazon	54	1,440±70	A.D. 543–551	0.049	Beta
				557–660	0.951	143615
Manacapuru	Central Amazon		1,525±58	A.D. 440–451	0.074	P406
				465–504	0.263	
				506–518	0.072	
				528–601	0.591	
Oswaldo	Central Amazon	41	1,550±40	A.D. 436–523	0.831	Beta
				526–542	0.161	143608
				555–555	0.008	
Pedrera	Japurá/Caquetá		1,390±210	A.D. 433–784	0.840	GrN8459
				788–832	0.088	
				837–876	0.072	
Mangueiras AM-JA-3	Japurá/Caquetá	Cut 2, 0–15 cm	1,525±58	A.D. 425–641	1.000	P409
Oswaldo	Central Amazon	34	1,730±90	A.D. 182–187	0.012	Beta
				216–421	0.988	143611
Itacoatiara	Central Amazon		1,864±58	A.D. 82–111	0.199	P372
				115–223	0.801	
Itacoatiara	Central Amazon		1,855±150	B.C. 17–14	0.006	
				A.D. 1–343	0.983	
				372–378	0.010	
Oswaldo	Central Amazon	80–90	1,980±80	B.C. 35–34	0.012	Beta
				18–13	0.053	143621
				A.D. 1–34	0.548	
				36–60	0.387	
Oswaldo	Central Amazon	90–100	2,120±40	B.C. 199–186	0.095	Beta
				B.C. 183–91	0.829	143622
				72–62	0.076	

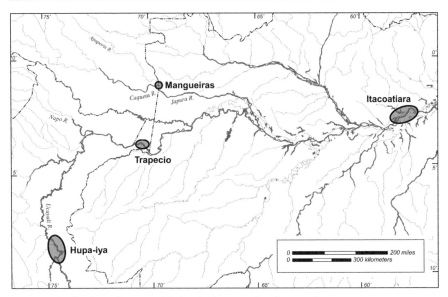

Fig. 6.6. Distribution of Barrancoid sites in the central and upper Amazon

River, near the mouth of the Apaporis River at the border between Brazil and Colombia, in about A.D. 400. On the lower Madeira, the earliest dates that are attributable to Barrancoid peoples are after A.D. 800 (Table 6.1). While the existing dates might seem to suggest that Barrancoid peoples originated on the Ucayali, too few sites have been radiocarbon dated to suggest that the earliest radiocarbon dates mark the beginning of the Barrancoid occupation in any particular region. Distributional evidence suggests that the Negro or Orinoco Rivers are the most likely homeland of the Barrancoid tradition (Lathrap 1970).

6.4.1
Barrancoid Sites on the Japurá and Caquetá Rivers

There are a number of Barrancoid sites on the Japurá and Caquetá Rivers around the mouth of the Apaporis River, which marks the border between Colombia and Peru, as well as on affluents of the Apaporis. The Mangueiras site, on the Brazilian side of the border below the Apaporis River, seems to be the largest of the group. Two test pits, separated by a distance of 60 m, had similar stratigraphy. Two levels of high sherd concentration are separated by 15–30 cm of lower density in each of the pits. The upper level is probably associated primarily with sherds of the Polychrome tradition, and the lower level with remnants of the Barrancoid tradition. In both pits the uppermost cultural level is associated with dark brown earth that becomes mottled with increasing amounts of yellow clay the deeper it gets. Charcoal is present, but it is not sufficiently prominent that it is mentioned in the descriptions of the soil. Two calibrated radiocarbon dates from the lower sherd level, one from each of the pits, suggest that the site was occupied between A.D. 500 and 700 (Hilbert 1968; Table 6.2).

Barrancoid sites on the Colombian side of the border, above the Puerto Córdoba Rapids, seem to be somewhat smaller, measuring just 30 m in diameter and having depths of 40–100 cm. The soils are not described, but charcoal is present. Two radiocarbon dates of A.D. 560 probably date the Barrancoid occupation, while a date of A.D. 1110 probably pertains to an occupation of the Polychrome tradition (Reichel 1984, 1988; Reichel and Hildebrand 1983; Herrera 1987).

None of the Barrancoid sites on the Japurá and Caquetá Rivers seems to be very large. A site that is 30 m in diameter may be compared to *malocas* on the Tiquié and Aiary Rivers mentioned above, some of them only 150 km away. Nevertheless, an accumulation of 40–100 cm of compost over an area that is 30 m in diameter suggests that intensive agriculture had been practiced over an extended period of time, a fact that argues for both population size and density.

6.4.2
Barrancoid Sites on the Ucayali River

Barrancoid sites on the Ucayali stretch from the mouth of the Pachitea River (Roe 1973) to Pirococha near the modern town of Contamana (Myers 1967). Nevertheless, the Barrancoid occupation of the central Ucayali was short-lived. By A.D. 90, makers of the Yarinacocha Complex had replaced it.

There are at least two large and important sites of the Barrancoid tradition, locally called the Hupa-iya Complex, on the central Ucayali near Pucallpa – UCA 1 and 2. Also there was a smaller occupation at UCA 34, across the ravine from UCA 6. Unfortunately there are no radiocarbon dates for the Hupa-iya Complex, but on the basis of a single date of A.D. 90 (uncalibrated) for the succeeding Yarinacocha Complex, Lathrap suggested a date of about 300 B.C. for Hupa-iya, somewhat earlier than radiocarbon dates from the central Amazon, but those dates may not date the earlier portions of the Barrancoid presence in that region.

Lathrap estimated that UCA 2 extended along the bluff for a distance of some 500 m, but was only about 60 m in width, an area of about 3 ha. Over most of its extent the midden is quite shallow, but in at least two areas it reaches a depth of 1.75 m. Lathrap believed that there were at least six, and perhaps seven or eight distinct occupations beginning some 4,000 years ago. The Hupa-iya occupation covered an area of at least 1.5 ha, the distance between productive excavations, but may have been much larger. In one area of the site the cultural deposit reached a depth of 80 cm, most of it attributable to the Hupa-iya occupation (Lathrap 1962).

In 2001, a layer of black earth, 10–15 cm thick, was visible where the original ground surface has been preserved along a fence line at the eastern end of the site not far from Lathrap's excavations. There is a similar layer of black earth at the northwestern corner of the football field close to Lathrap's first excavation nearly 50 years ago. Several Pacacocha sherds were eroding from the surface of the midden. Cultural materials were concentrated in the upper-most 15 cm of Lathrap's test pits, although excavations continued to a depth of 30 cm. The soil was mainly clay, though darkened with charcoal in the uppermost 15 cm. Excavation produced a few Shipibo sherds, quite a few Pacacocha sherds, and some Hupa-iya material, but the midden was not deep enough to provide stratigraphic separation (Lathrap 1958, 1962). Elsewhere on the site Lathrap's excavations seem to have taken place in occupied areas where the soil line of black earth had already been removed.

From the standpoint of cultural stratigraphy, Cut 3 was the most important at the site and the only one that was described in detail. The uppermost layer was an active Shipibo midden that was still accumulating at the outer edge of the plaza. It consisted of a dark gray to black sandy clay that included char-coal and ash as well as shell and bone. The second layer was the Hupa-iya occupation level. It consisted of light gray, sandy clay with no visible char-coal. The lack of charcoal contrasts with the abundance of charcoal in dark earth sites of the central and lower Amazon. Cultural material was sparse at the top but denser toward the bottom of the stratum. Many years later Lath-rap sometimes thought of this as a *terra preta* site (Eden et al 1984). The third level at UCA 2 was attributed to the Shakimu occupation. Lathrap described it as a thick band of bright orange sandy clay with very heavy concentrations of baked clay fragments. The fourth stratum was a red brown layer of sandy clay containing scattered cultural material of the Tutishcainyo Complex. The fifth stratum was a thick bed of sand, nearly barren of cultural material. Lath-rap noted that almost every trace of organic material had been leached out of all pre-Shipibo deposits. Even occasional flecks of charcoal and charcoal stains were rare (Lathrap 1962). In comparison with Barrancoid sites else-where, the absence of charcoal in the midden at UCA 2 is remarkable. In spite of Lathrap's later opinion it is difficult to think of this as a black earth site.

6.5
Polychrome Migrations Out of the Central Amazon

The Polychrome tradition developed out of the Barrancoid tradition on the central Amazon by about A.D. 500 or 600 (Lathrap 1970). Within a few hun-dred years, Polychrome-making peoples had begun to migrate upstream, passing the Trapecio de Amazonas, where Colombian soil touches the main-stream of the Amazon, shortly after A.D. 1000. They reached the mouth of the Apaporis on the Japurá River shortly after A.D. 1100 and as far as the Arara-cuara Plateau by the mid 1200s. They ascended the Napo River as far as the mouth of the Aguarico a little before A.D. 1200, and the Ucayali as far as the Pachitea by shortly before A.D. 1300 (Fig. 6.7). In every case, the Polychrome occupation followed the Barrancoid occupation by several hundred years. In

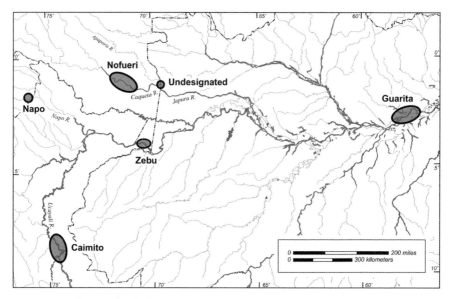

Fig. 6.7. Distribution of Polychrome sites in the central and upper Amazon

the Peruvian Amazon the Polychrome tradition is associated with the Tupi-speaking Omagua and Cocama. Tupi speakers were probably also the bearers of this tradition into the Japurá and Madeira River drainages (Brochado 1984).

If both Lathrap and Neves and associates are correct, the Polychrome and Barrancoid traditions flourished side-by-side for several hundred years on the central Amazon. The data necessary to assess this possibility have not yet been published. Fortunately, for the purposes of this chapter, these are not important. Polychrome occupations clearly follow Barrancoid occupations on the Japurá and Ucayali Rivers, and probably on the Madeira River as well.

6.5.1
Polychrome Sites on the Japurá and Caquetá Rivers

There are a group of Polychrome tradition sites on both sides of the border, above and below the mouth of the Apaporis River. Many sites had both Barrancoid and Polychrome occupations, but none of the materials has been reported in detail. At the Mangueiras site the Polychrome occupation is attributable to the upper sherd level in which the soil is described as uniformly dark brown earth. A radiocarbon date from the Colombian side of the border suggests a calibrated date of shortly after A.D. 900 (Herrera 1987; Table 6.2).

There is a second group of dark earth Polychrome tradition sites, the Nofueri Phase, upstream near Araracuara, on the Caquetá River, which had

Table 6.2. Calibrated radiocarbon dates (1δ) on the Japurá and Caquetá Rivers. Calibrated using CALIB v4.3. (After Hilbert 1968; Herrera 1987)

Site/cut/ depth	Phase (Herrera)	Sample no.	Radio-carbon age (years B.P.)	Calibrated age	Proba-bility	Region	Tradition (Myers)
Sitio 25/2	Camani	Beta 6949	2,740±90	B.C. 996–990	0.027	Arara-cuara	Local
				974–949	0.132		
				945–808	0.842		
ARA 21	Camani	Beta 1503	1,815±105	A.D. 82–263	0.763	Arara-cuara	Local
				276–338	0.237		
ARA 15	Camani	IAN 113	1,800±85	A.D. 95–95	0.004	Arara-cuara	Local
				127–263	0.701		
				274–339	0.295		
ARA 22	Camani	Beta 1504	1,690±65	A.D. 257–284	0.178	Arara-cuara	Local
				286–301	0.085		
				319–423	0.737		
ARA 15	Camani	Beta 1509	1,480±95	A.D. 441–450	0.040	Arara-cuara	Local
				466–484	0.081		
				510–517	0.030		
				529–657	0.797		
ARA 15	Camani	Beta 1505	1,420±70	A.D. 545–546	0.006	Arara-cuara	Local
				560–671	0.994		
AM-JA-3 Cut 2, 0–15 cm	Manguei-ras	P409	1,525±58	A.D. 425–641	1.000	Japurá	Barrancoid
No data		GrN8459	1,390±210	A.D. 433–784	0.840	La Pedrera	Barrancoid
				788–832	0.088		
				837–876	0.072		
No data		SI6375	1,390±75	A.D. 564–571	0.036	La Pedrera	Barrancoid
				578–588	0.049		
				597–692	0.815		
				701–712	0.057		
				751–761	0.043		
AM-JA-3 Cut 1, 60–75 cm	Manguei-ras	P588	1,315±59	A.D. 622–628	0.006	Japurá	Barrancoid
				638–784	0.873		
				788–832	0.069		
				837–876	0.052		
ARA 26	Camani	Beta 6950	1,160±50	A.D. 782–791	0.073	Arara-cuara	Local
				809–844	0.242		
				850–900	0.374		
				919–959	0.311		

Table 6.2. (Continued)

Site/cut/ depth	Phase (Herrera)	Sample no.	Radio-carbon age (years B.P.)	Calibrated age	Proba-bility	Region	Tradition (Myers)
ARA 15	Camani	Beta 1507	1,145±80	A.D. 782–791	0.054	Arara-cuara	Local
				809–845	0.183		
				849–980	0.763		
ARA 15	Camani	Beta 1508	1,120±65	A.D. 783–788	0.025	Arara-cuara	Local
				830–838	0.038		
				870–998	0.937		
		SI 6374	840±40	A.D. 1163–1173	0.098	La Pedrera	Poly-chrome
				1180–1256	0.902		
ARA 15	Nofurei	Beta 1506	705±60	A.D. 1259–1313	0.670	Arara-cuara	Poly-chrome
				1354–1387	0.330		
ARA 7	Nofurei	Beta 1510	350±50	A.D. 1489–1528	0.318	Arara-cuara	Poly-chrome
				1550–1633	0.682		

been occupied by a local tradition during Barrancoid times. A number of small sites identified within the town of Araracuara are likely to be the remnants of one large site (ARA 15) that stretched for a distance of 2 km, but was only 300 m wide. Other Polychrome sites (ARA 7, 20) are also on the floodplain, some on top of Camani Phase sites. Radiocarbon dates place the Polychrome occupation between A.D. 1245 and 1610, uncalibrated. Several other Polychrome sites covering 2 ha or more were also found on the floodplain (Herrera et al. 1981; Eden et al. 1984; Andrade 1986; Herrera 1987). Another black earth site of the Nofueri Phase, the Sardina site (ARA 20), is located on the floodplain.

6.5.2
Polychrome Sites on the Napo River

Polychrome sites on the Napo River are identified as the Napo Phase with radiocarbon dates between A.D. 1168 and 1480. They are clearly attributable to the Tupian occupation historically known as the Omagua-yete on this part of the Napo. Sites recorded by Evans and Meggers are both small and remarkably shallow. Often the remaining midden is no more than 5 cm deep. Midden soils are usually blackish- to grayish-brown sandy loam. There was a sufficient quantity of charcoal for radiocarbon dates even though charcoal was not mentioned in the description of the soil (Evans and Meggers 1968).

6.5.3
Polychrome Sites on the Ucayali River

The Polychrome presence on the central Ucayali, locally called the Caimito complex, is best known from sites on the Tamaya River, an eastern tributary of the Ucayali that enters some distance below the mouth of the Pachitea River. Most of the sites are located on the alluvial lands next to Imariacocha, a dammed lake remarkable for its many inlets. Sites range from 23–150 cm in depth, but in the latter case the occupation level is overlain by more recent alluvial deposits. The occupation levels are described as bright orange and tan mottled gumbo, tan to buff sandy loam, sandy brown clay, and reddish-brown sandy soil (Lathrap 1964; Weber 1975). In spite of Lathrap's later opinion (Eden et al. 1984), it is difficult to think of these as *terra preta* sites (Table 6.3).

Table 6.3. Natural stratigraphy of sites on the Tamaya River

Site	Depth	Location	Description of soils and cultural associations
TAM 1A	38 cm	*Terra firme*	36-ft. trench toward the lake. Four natural stratigraphic units: (1) (upper) tan sandy loam; (2) pale orange to orange-brown gumbo; (3) bright orange and tan mottled gumbo (occupation level); (4) yellow gumbo (Lathrap 1964)
TAM 2A	23 cm	*Terra firme*	(1) Tan to buff sandy loam with sherds (6 in.); (2) culturally sterile bright vermillion laterite clay. (Lathrap 1964)
TAM 2C	23 cm	*Terra firme*	Upper level is red laterite; lower (occupation) level is sandy brown clay. Excavations to a depth of 9 in. (Lathrap 1964)
Tam 11 (Shebonasi)		*Terra firme*	Although horizontal stratigraphy could not be defined, there was a gradient from a reddish-brown sandy textured soil, characteristic of cultural deposits, to a reddish laterite clay (Weber 1975)
Tam 11C (island)	1.5m	*Terra firme*	A layer of black humus at the surface overlies reddish-brown clay, which covers the cultural deposit described as gray-brown clay Cultural deposit ranges from 0.3–0.85 m in thickness (Weber 1975)
Tam 11D (Shebonasi)	50 cm	*Terra firme*	Top 5–20 cm made up of a black soil horizon associated with modern Conibo occupation. Below this level soil graded from reddish-brown sandy soil which contained the Caimito occupation to culturally sterile red laterite clay (Weber 1975)

Anthropogenic soils may not be entirely lacking from Imariacocha. An island designated TAM 11C produced evidence of a sparse Caimito occupation. This island, and another in the lake, may have been artificially created, or improved, for agricultural purposes by the Caimito people who brought in soil to raise the surface of the island by nearly 1 m. The original height of the fill may have been preserved by a house, while sheet erosion in the plaza caused deflation of the surrounding surface (Weber 1975).

6.6
Dark Earth and the Indigenous Peoples of the Japurá, Ucayali, and Napo Rivers

When Barrancoid and Polychrome peoples expanded out of the central Amazon they followed the major rivers, pushing native peoples aside and forcing them upstream or into the hinterland between the major rivers. Lathrap supposed that the native peoples, represented by local archaeological traditions (Fig. 6.8), occupied the major rivers before the arrival of invading Barrancoid or Polychrome peoples (Lathrap 1968). This is not necessarily the case. Cipolletti believes that the native peoples of the Napo River occupied the hinterland in prehistoric times, making only occasional use of the *várzea*, which was the primary habitat of the Napo Phase (Cipolletti 1997). The conflicting hypotheses can be tested with the archaeological record.

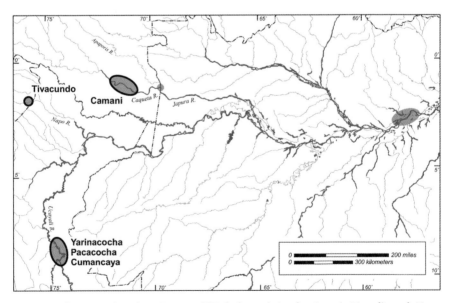

Fig. 6.8. Indigenous sites (non-Barrancoid/Polychrome) in the Japurá, Ucayali, and Napo basins

6.6.1
Indigenous Peoples of the Japurá/Caquetá

During the time of the Barrancoid occupation at the juncture of the Apaporis and Caquetá Rivers, the Araracuara region, some 250 km up the Caquetá, was occupied by people of the Camani Phase, part of a local tradition unrelated to either the Barrancoid or the Polychrome traditions. It seems to have dominated the Araracuara region from at least 800 B.C. until the arrival of the Polychrome tradition in the mid 1200s. Sites were located both on the banks of the Caquetá itself and in defensive positions on top of the Araracuara mesa which overlooks the Caquetá (Herrera et al. 1981; Eden et al. 1984; Andrade 1986; Herrera 1987; Mora 2004).

Soil enrichment activities took place during the Camani Phase, long before the appearance of the Polychrome tradition in the mid 1200s, and even before Barrancoid peoples appeared at the mouth of the Apaporis. Several dark earth sites are associated with the Camani Phase. The Aeropuerto site (ARA 3/4) is a dark earth site with black and brown soils on the northern flank of the Araracuara Plateau. *Terra preta* is found over an area of 6 ha, while dark brown earth covers an additional 20 ha. The soil is described as sandy loam with an average depth of 40 cm to 1 m. The site seems to have been occupied continually or intermittently over a period of some 2,000 years. Two radiocarbon dates range from 790 B.C. to A.D. 790, but the abundance of Nofueri Phase (Polychrome) pottery suggests that it was occupied considerably after this date. This is the type site for the Nofueri Phase (Herrera et al. 1981; Eden et al. 1984; Andrade 1986; Herrera 1987; Mora 2004).

The Abeja site is located in a depression on top of the Araracuara plateau. It covers an area of 6 ha with brown loamy-sand soils with an average depth of 22 cm. There are six radiocarbon dates from this site, ranging from A.D. 385 to 1176 (uncalibrated) (Mora 2004). The Sardina site (ARA 20), a floodplain black earth site of the Nofueri Phase, was initially occupied during the Camani Phase (Herrera et al. 1981; Eden et al. 1984). In contrast, there is no evidence of dark earth soils at the Puerto Arturo site located next to the river at the northern tip of the Araracuara plateau (Mora 2004).

Important changes took place in the Camani Phase around A.D. 800. Mora suggests that there was an abandonment of the Puerto Arturo and Abeja sites. At the same time, ceramics became more elaborate and carefully made. Humidity indicators in the palynological sample from the Aeropuerto site suggest that sediments (algae) imported from the lower-lying areas may have become part of the anthropic soil formation process, although Mora suggests that more data are needed to evaluate this hypothesis (Herrera et al. 1992; Mora 2004).

The changes that took place around A.D. 800, and in particular the concentration into larger sites on top of the mesa, may have been a response to a threat posed by the appearance of Barrancoid peoples downstream. If so, we may be witnessing the formation of a secondary chiefdom, which appears as a response to a threat posed by the appearance of a primary chiefdom (Service 1962).

6.6.2
Indigenous Peoples of the Ucayali

Prior to the arrival of Barrancoid peoples, the Ucayali was occupied by peoples of the Shakimu complex, a late component of the Tutishcainyo tradition, which dominated the region for 2,000 years before the arrival of Barrancoid peoples. Following the short-lived Hupa-iya occupation were a succession of local traditions designated Yarinacocha, Pacacocha, and Cumancaya, respectively. Peoples of the Polychrome tradition ascended as far as the Pachitea by the late 1200s but soon retreated to the lower Ucayali. They were displaced by late components of the Cumancaya tradition that included both Pano and Arawak speakers in historic times.

6.6.2.1
Tutishcainyo Tradition

There is no evidence of dark earth associated with the Tutishcainyo tradition. Soils associated with this tradition are described as red-brown sandy clay at UCA 2 and as dark brown sandy clay at UCA 6 (Lathrap 1962). The paucity of charcoal is emphasized by Lathrap's lament that time prevented him from excavating a rounded oval depression crammed with Shakimu pottery and numerous minute charcoal flecks at UCA 2 in the last days of the field season (Lathrap 1962).

6.6.2.2
Yarinacocha Tradition

The Yarinacocha complex followed Hupa-iya on the central Ucayali shortly before A.D. 100. Known from only two sites, located across the ravine from one another, it is distinguished by a wide range of vessel forms and polychrome painted decoration, though it is clearly not a member of the Polychrome tradition. The Yarinacocha complex is best represented at UCA 34 where a large series of trash pits filled with pottery were placed on the slope of the hill. Excavations in an overgrown garden on top of the hill, where the village must have been located, revealed 10 cm of dark brown sandy clay, then 18 cm of light brown sandy clay overlying red laterite. Refuse was sparse (Myers' field notes, 1964). This pattern of refuse distribution conforms to the model anticipated from a comparison of modern village sites (see above; Myers 1973).

6.6.2.3
Pacacocha Tradition

Pacacocha phase sites on the central Ucayali are generally characterized by a thin layer of black to gray humus or sandy loam followed by a lighter colored soil, usually somewhat sandy, and then laterite (Table 6.4). UCA 10 was unusual in that it was a much deeper site with a somewhat greater number of sherds, and several cultural components. At UCA 10 the greater part of the cultural materials come from a soil described as gray sandy loam (Table 6.5).

Table 6.4. Summary of natural stratigraphy of Pacacocha Phase sites at Yarinacocha. (After Myers 1970)

Site	Location	Description of soils
UCA 4	*Terra firme*	Upper soil level was characterized by an inch or two of black humus followed by a layer of brown loam mixed with sand, and finally laterite clay
UCA 10	*Terra firme*	About 12 in. of gray sandy loam overlay about 2 ft. of brown sandy loam which, in turn, overlay red laterite clay
UCA 17	*Terra firme*	Thin layer of dark brown humus mixed with charcoal from recent burnings formed the upper soil zone. This was followed by a zone of light brown sandy loam, and finally red laterite clay
UCA 33	*Terra firme*	Top 3–4 in. in each pit were composed of a tan sandy clay. Below this level was red laterite clay

Table 6.5. Natural and cultural stratigraphy (sherd counts) at UCA 10

Depth (in.)	Natural stratigraphy	Tutishcainyo	Cashibocaño	Cumancaya	Shipibo
0–3	Gray sandy loam		354	163	230
3–6			219	154	6
6–9			284	19	
9–12			679	8	
12–15	Brown sandy loam		493	2	
15–18			117		
18–21			38		
21–24		1	94[a]		
24–27		5	31		
27–30		4	59		
30–33			36		
33–36			9		
36–39			5		

[a] Large number of sherds at top of feature that extended into next level.

However, a significant number also come from the clay described as brown sandy loam, gradually thinning out as the underlying red laterite soil is approached.

Soil colors mentioned in the description of Pacacocha sites include black humus and brown loam at the surface, followed by gray sandy loam or brown sandy loam. The level of organic content is seldom very high. Photographs of the main trench at UCA 17 suggest that the soils might better be described as gray sandy soils for the entire depth of the cultural deposit. However, the near absence of charcoal suggests that they are not *terra preta*.

6.6.2.4
Cumancaya Tradition

Sites of the Cumancaya tradition are best known from the upper Ucayali, above the mouth of the Pachitea River, and from the Tamaya River, which enters the Ucayali just below the mouth of the Pachitea from the opposite side. The Cumancaya site (UCA 22) is a multicomponent site on the flood-plain less than 1 m above the level of the adjacent oxbow lake. Hupa-iya, Cumancaya, and Caimito occupations overlie one another in a midden that is just 20–30 cm thick. A late Shakimu component was present elsewhere on the site. The midden layer is described as being dark brown clay filled with sherds and flecks of carbon that might qualify as dark earth. It was overlain by three soil zones to a depth of 40–50 cm: (1) a layer of forest lit-ter; (2) a layer of hard light brown clay; and (3) a layer of hard, yellow brown clay. Dense culturally sterile clay underlies the midden (Raymond et al. 1975).

The Sonochonea site (UCA 40) is also located in the floodplain of the Ucayali, some 2 m above the level of the river in May 1971. The occupation level of brown silt is covered by several layers of alluvium beginning with gray and yellow mottled clay, brown clay loam, brown sandy loam and white sandy loam at the surface, 5 m above the level of the water. The single occupa-tion period has a radiocarbon date of A.D. 830 (Raymond et al. 1975).

The Shahuaya site (SHA 1) is located on *terra firme* overlooking the Sha-huaya River, a small western tributary of the Ucayali. The midden begins at a depth of about 40 cm. It is about 30 cm thick, composed of brown, clayey loam with abundant pottery. A radiocarbon date of A.D. 1630 is believed to date the occupation of the site, midway between the initial exploration of the Ucayali and the Jesuit mission period. Overlying the midden is a layer of yel-low sandy loam to a depth of about 40 cm. Beneath the midden is a sandy deposit, which, in turn, overlies a layer of yellow sand, and finally a layer of white sand (Raymond et al. 1975).

The Granja de Sivia site is on a terrace of the Apurimac River, far to the south, well beyond the advance of Polychrome peoples. A series of radiocar-bon dates suggest that the site was occupied from about A.D. 950 to 1300. The dark brown midden varies from 30 to 120 cm in thickness over an area of at least 2 ha. The midden layer continues into the thin layer of sod at the surface of the site. Beneath the midden is light brown sterile sand (Raymond et al. 1975).

6.6.2.5
Local Traditions on the Upper Pachitea

The archaeological record is quite different on the upper Pachitea River where most desirable hilltop locations were capped by multicomponent sites with deep stratigraphy. All components over a period of nearly 4,000 years are attributed to a local tradition, the Cobichiniqui tradition, with the exception of an intrusion of the Cumancaya-like Naneni complex about A.D. 600. There is no evidence of Barrancoid or Polychrome occupations of the region (Allen 1969).

Excavations at the Casa de la Tía site, on the Nazaratequi River, revealed cultural materials to a depth of more than 1.5 m. Cut D exhibits a clear association between natural and cultural stratigraphy (Table 6.6). Two zones might be construed as *terra preta*. The uppermost, which occurred between 9 and 17 in. (23–43 cm), was described as "very dark black humus." It is associated

Table 6.6. Natural and cultural stratigraphy (sherd counts) of Cut D at the Casa de la Tia site (PAC 14). (After Allen 1968)

Depth (in.)	Natural stratigraphy	Radiocarbon dates	Cobi-chiniqui Complex	Pangotsi Complex	Nazaratequi Complex	Enoqui Complex
0–3	Loosely consoli-				11	17
3–6	dated silt, ash, and				47	150
6–9	sand				262	244
9–12	Very dark black				1,018	239
12–15	humus				1,642	177
15–18		A.D. 604±69			2,788	84
18–21					699	30
21–24					283	
24–27					180	
27–30	Brown humus-			25	9	
30–33	midden, rock,	670±100 B.C.		54	6	
33–36	sand			40	6	
36–39		1275±68 B.C.		100	9	
39–42			1	60	2	
42–45		1637±95 B.C.	15	42		
45–48	Dark black, comp-		11	8		
48–51	acted midden		9			
51–54		1778±65 B.C.	9			
54–57			11			
57–60		1418±77 B.C.	11			

with the Nazaratequi occupation, about A.D. 600. Allen estimated that a thousand people lived on the Casa de la Tía site during the Nazaratequi period. The lower, between 45 and 63 inches (114–160 cm), was described as "dark black compacted midden." It is associated with the Cobichiniqui complex dating to 1500 B.C., or earlier.

6.6.3
Indigenous Peoples of the Napo River

Both archaeological evidence and linguistic distributions north of the Amazon on the Napo and Putumayo Rivers imply a different historical development than is found on the Ucayali. In this region north of the Amazon, a more or less solid block of western Tucanoans is subdivided by Tupians living along the major rivers. In fact, there were not a great many Tupians. When Orellana descended the Napo he found Omagua only on the upper Napo. Below the Aguarico the banks of the Napo were uninhabited until he got very close to the Amazon where he again found Omagua speakers. Cipolletti is eloquent in her demonstration that the western Tucanoans were adapted to the *terra firme* and not to the *várzea* (Cipolletti 1997).

The Tivacundo Phase, which preceded the Napo Phase of the Polychrome tradition on the Napo River where it is joined by the Aguarico, is probably attributable to western Tucanoan peoples. There is a radiocarbon date of A.D. 510 (uncalibrated). Only two sites are known, both adjacent to the mainstream. One is the tiny remnant of a site that had mostly washed away. The other site, Chacra Alfaro, was just 30–35 m in diameter, covering an area of somewhat less than 1 ha. The occupation layer was about 10 cm thick. The soils of this layer are described as being brownish tan in color, overlying the brown loamy clay of the alluvial soils. There was enough charcoal to provide a radiocarbon date, but the brownish tan color of the soil does not suggest *terra preta* (Evans and Meggers 1968).

Though separated by 1,000 years, peoples of the Tivacundo Phase may have been the ancestors of the Encabellado and Abixira peoples who occupied the region in the seventeenth century. The fact that one or two sites were close to the Napo River 500 years before the arrival of Polychrome immigrants does not necessarily controvert Cipolletti's contra-hypothesis, but it does not support it either. The area covered by the site corresponds to the area that might be expected from a single *maloca* or multifamily dwelling of the sort used by the earliest historic inhabitants of the region.

6.7
Conclusions

The ethnographic evidence suggests that the habitation areas of sites are regularly kept clean. They are swept. Trash is picked up and cast into the brush along with human or animal excrement. In many villages the surface of the site is being degraded as it is at San Francisco de Yarinacocha rather than being allowed to accumulate. This manner of housekeeping makes it unlikely that extensive areas of black earth would be formed in habitation areas. Instead, black earth is likely to naturally form in areas of refuse disposal – in effect, a compost pile. Excavations at UCA 34 revealed precisely this pattern with a rather deep midden on the slope of the hill but virtually no refuse accumulation on top of the hill. Refuse distribution patterns at UCA 2 and UCA 6 could be interpreted in the same fashion. Thus explanation offered by Herrera and associates for the Araracuara region is most likely to be correct. Black earth was purposefully formed in composting areas outside of the habitation area, possibly in a kitchen garden between or behind houses. House floors and other evidence of habitation found in black earth sites are likely to have been abandoned before being covered with compost.

Black earth is regularly associated with the earliest Barrancoid sites of the central Amazon. It is also associated with Barrancoid sites on the Japurá River, but evidently not on the Ucayali. Similarly, Polychrome sites of the central Amazon are regularly associated with black earth, but Polychrome sites on the Ucayali and Napo Rivers are not except for the islands of Imariacocha. In contrast, Polychrome sites on the Caquetá River continue the pattern of black earth production established prior to the Barrancoid intrusion.

The Polychrome occupation of both the central Ucayali and the Napo was relatively short. On the Ucayali, indigenous peoples seem to have driven the invaders downstream, away from Imariacocha, well before the arrival of Europeans, leaving a period of less than 200 years, probably closer to 100, for the development of black earth. If black earth is the product of composting, the time may have been too short for large areas of black earth to be produced and distributed.

The Polychrome occupation of the Napo River was considerably longer, from about A.D. 1200 to the arrival of the Spaniards in 1540, a period of more than 300 years. Yet there is no evidence of black earth being associated with sites of the Napo Phase. Sites seem to be relatively small and short-lived when compared to Polychrome tradition sites elsewhere. Middens are usually blackish to grayish brown sandy loam. While there is enough charcoal to produce some radiocarbon dates, there seems to be little resemblance to the dark, highly fertile soils that characterize black earth sites of the central Amazon. Possibly because the indigenous Tucanoan peoples posed little threat, the invading Tupians found little need to maintain large sites or to invest the labor necessary for the production of black earth fields. They were able to switch to an extensive form of agriculture just as mixed bloods (*caboclos* or *mestizos*) were able to do in conditions of low population density following the rubber boom.

In general, the soils of archaeological middens on the Ucayali are light-colored and shallow compared with those in Brazil. There are two notable exceptions: one at the Casa de la Tía site on the upper Pachitea, the other at the Cashibocaño site at Yarinacocha. The fact that they are so very unusual suggests that they are not part of an ongoing tradition of black earth production. They must have been produced under very special circumstances. Though both sites are on top of a hill or bluff, the excavations were made on a slope that may have been an area of refuse accumulation over an extended period of time.

Conversely, the sites of both indigenous peoples and immigrants on the Japurá/Caquetá River are regularly black earth sites. There is some evidence to suggest that black earth technology improved by the addition of imported algae around A.D. 800 in the face of a threat from Barrancoid peoples living some 250 km downstream. The addition of algae may or may not have been a local innovation.

In conclusion, given the sanitation practices of Amazonian peoples, black earth associated with habitation sites is most likely to be a product of composting followed by the purposeful spread of compost over abandoned habitation areas. While black earth technology may have been an essential component of the aboriginal adaptation to the Amazonian heartland, it was not limited to that region. It was practiced on the Caquetá River long before the arrival of Barrancoid intruders. In contrast, there is little to suggest that black earth technology was utilized by Polychrome immigrants to the Napo River who did not need large, dense populations to control the *várzea*. Barrancoid intruders on the Ucayali may not have had the opportunity to develop black earth technology since they were soon driven away by local peoples. The same fate awaited Polychrome invaders some 1,500 years later. They had only begun to develop black earth sites on islands in Imariacocha before they were forced back by peoples of the Cumancaya tradition.

References

Allen WL (1969) A ceramic sequence from the Ucayali River, Peru: some implications for the development of tropical forest culture in South America. PhD Diss, University of Illinois, Urbana

Andrade A (1986) Investigaciones Arqueológicas de los Antrosoles de Araracuara. Fundación de Investigaciones Arqueológicas Nacionales, Banco de la República, Bogotá

Blomberg R (ed) (1952) Ecuador: Andean mosaic. H Geber, Stockholm

Brochado JJ (1984) An ecological model of the spread of pottery into eastern South America. University Microfilms, Ann Arbor

Carneiro R (1970) A theory of the origin of the state. Science 169:733–738

Castrucci VM (1905) Viaje del párroco de Andoas, frai Manuel Castrucci Vernozza, a los territorios habitados por los Zaparos i Jibaros en los rios Pastaza, Napo i Bobonaza (1849). In: Larrabure i Correa C (ed) Colección de leyes, decretos, resoluciones i otros documentos oficials referentes al Departamento de Loreto, vol 6. Imprenta de la Opinión Nacional, Lima, pp 508–541

Cipolletti MS (1997) Los Tucano del Alto Amazonas. Un contramodelo al modelo de dinámica poblacional de Lathrap. In: Cipolletti MS (ed) En Resistencia y adaptación en las tierras bajas latinoamerias. Colección Biblio AbyaYala 36:323–342

Cruz de la L (1999) Descripción de los Reynos del Perú con particular noticia de lo hecho por los Franciscanos. Pontificia Universidad Católica del Perú by Banco Central de Reserva del Perú, Lima.

Domville-Fife CW (1924) Among wild tribes of the Amazons. Seeley, Service and Co, London

Eden MJ, Warwick B, Leonor H, Colin ME (1984) *Terra preta* soils and their archaeological context in the Caquetá basin of southeastern Colombia. Am Antiq 49(1):124–140

Evans C, Meggers BJ (1968) Archeological investigations on the Rio Napo, eastern Ecuador. Smithson Contrib Anthropol 6

Forge A (1972) Normative factors in the settlement size of neolithic cultivators (New Guinea). In: Tringham R (ed) Ecology and agricultural settlements: an ethnographic and archaeological perspective. Warner Module, Andover, Massachusetts

Goldman I (1963) The Cubeo: Indians of the northwest Amazon. Illinois Studies in Anthropology 2. University of Illinois Press, Urbana

González EM, Tur NE (eds) Publicaciones y Ediciones de la Universidad de Barcelona, Editorial 7 1/2. Universidad de Barcelona, Barcelona, pp 203–271

Heckenberger MJ (1996) War and peace in the shadow of empire: sociopolitical change in the upper Xingu of southeastern Amazonia, ca. A.D. 1400–2000. University Microfilms, Ann Arbor

Hernandez C (1981) Relación muy berdadera que trata de todo lo que acaeció en la entrada de Pedro de Orsua. In: Lope de Aguirre C (1559–1561), González EM, Tur NE (eds) Publicaciones y Ediciones de la Universidad de Barcelona, Editorial 7 1/2. Universidad de Barcelona, Barcelona, pp 191–200

Herrera LF (1987) Apuntes sobre el estado de la investigación arqueológica en la Amazonía colombiana. Bol Antropol 6(21):21–61

Herrera LF, Bray W, McEwan C (1981) Datos sobre la Arqueología de Araracuara (Comisaría del Amazonas Colombia). Rev Colomb Antropol 23:185–251

Herrera LF, Cavelier L, Rodriguez C, Mora S (1992) The technical transformation of an agricultural system in the Colombian Amazon. World Archaeol 24:98–113

Hilbert PP (1968) Archäologische Untersuchungen am mittleren Amazonas: Beiträge zur Vorgeschichte des südamerikanischen Tieflandes. Dietrich Reimer, Berlin

Hilbert PP, Hilbert K (1980) Resultados preliminares da pesquisa arqueológica nos rios Nhamundá e Trombetas, Baixo Amazonas. Bol Mus Paraense Emílio Goeldi Nova Ser Antropol 75

Kern DC, D'Aquino G, Rodrigues TE, Lima Frazão FJ, Sombroek W, Myers TP, Neves EG (2003) Distribution of Black Earths in the Brazilian Amazon. In: Lehmann J, Kern D, Glaser B, Woods WI (eds) Amazônian dark earths – origin, properties and management. Kluwer, Dordrecht, pp 51–76

Koch-Grünberg T (1910) Zwei Jahre unter den Indianern: Reisen in Nordwest-Brasilien 1903/1905. Strecker and Schröder, Stuttgart

Lathrap DW (1958) The cultural sequence at Yarinacocha, eastern Peru. Am Antiq 23(4):379–388

Lathrap DW (1962) Yarinacocha: stratigraphic excavatons in the Peruvian Montaña. PhD Diss, Department of Anthropology, Harvard University, Cambridge

Lathrap DW (1964) Field notes. On deposit in the Anthropology Division, University of Nebraska State Museum, Lincoln

Lathrap DW (1968) The "hunting" economies of the tropical forest zone of South America: an attempt at historical perspective. In: Lee RB, Devore I (eds) Man the hunter. University of Chicago Press, Chicago, pp 23–29

Lathrap DW (1970) The Upper Amazon. Praeger, New York

Mora S (2003) Archaeobiological methods for the study of Amazonian Dark Earths. In: Lehmann J, Kern D, Glaser B, Woods WI (eds) Amazônian dark earths – origin, properties and management. Kluwer, Dordrecht, pp 205–226

Myers TP (1967) Reconocimiento arqueológico en el Ucayali central. Bol Mus Nacl Antropol Arqueol 6

Myers TP (1970) The Late Prehistoric period at Yarinacocha, Peru. University Microfilms, Ann Arbor

Myers TP (1973) Toward the reconstruction of prehistoric community patterns in the Amazon Basin. In: Lathrap DW, Douglas J (eds) Variation in anthropology. Illinois Archaeological Survey, Urbana, pp 233–252

Myers TP (1992) Agricultural limitations of the Amazon in theory and practice. World Archaeol 24(1):82–97

Myers TP (2004) Cambios culturales y demográficos en el Solimões, 1542–1700. In: Cipolletti MS (ed) Los mundos de abajo y los mundos de arriba: Individuo y sociedad en las tierras bajas y en los Andes. Tomo de homenaje a Gerhard Baer. Abya Yala, Quito (in press)

Myers TP, Denevan WM, WinklerPrins AMGA, Porro A (2003) Historical perspectives on Amazonian Dark Earths. In: Lehmann J, Kern D, Glaser B, Woods WI (eds) Amazônian dark earths – origin, properties and management. Kluwer, Dordrecht, pp 15–28

Ortiguera de T (1981) Jornada del Marañón (1586). In: Lope de Aguirre C (1559–1561), González EM, Tur NE (eds) Publicaciones y Ediciones de la Universidad de Barcelona, Editorial 7 1/2. Universidad de Barcelona, Barcelona, pp 32–174

Raymond JS, DeBoer WR, Roe PG (1975) Cumancaya: a Peruvian ceramic tradition. Occasional Pap 2. Department of Anthropology, University of Calgary

Reichel DE (1984) Nota sobre investigaciones arqueologicas recientes sobre 'terra preta' en la Amazonia colombiana: Rio Cáqueta. Noticias antropológicas. Soc Antropol Colombia 82

Reichel DE (1988) Asentamientos prehispánicos en la Amazonia colombiana. In: Reichel DE (ed) Colombia Amazónica. Universidad Nacional de Colombia y Fondo 'Jose Celestino Mutis' FEN, Medellin, pp 129–144

Reichel Hildebrand E (1976) Resultados preliminares del reconocimiento del sitio arqueológico de la Pedrera (Comisaría del Amazonas, Colombia). Rev Colomb Antropol 20:145–177

Reichel E, von Hildebrand M (1983) Reconocimiento arqueologico del area del Bajo Rio Caquetá y Apaporis, Amazonas Noticias antropologicas. Soc Antropol Colomb 76/77:6–7

Roe PG (1973) Cumancaya: archaeological excavations and ethnographic analogy in the Peruvian Montaña. PhD Diss, University of Illinois, Urbana

Rowlett R (2003) How many? A dictionary of units of measurement. Center for Mathematics and Science Education University of North Carolina at Chapel Hill, http://www.unc.edu/~rowlett/units/index.html

Service ER (1962) Primitive social organization: an evolutionary perspective. Random House, New York

Simões MF, de Araujo-Costa F (1978) Areas da Amazonia Legal Brasileira para pesquisa e cadastro de sitios arqueológicos. Mus Para Emilio Goeldi Publ Avulsas 30

Simões MF, Corrêa C (1987) Pesquisas arqueológicas no baixo Uatumá-Jatapu (Amazonas). Rev Arqueol 4(1):29–48

Simões MF, Machado AL (1987) Pesquisas arqueológicas no lago de Silves (Amazonas). Rev Arqueol 4(1):49–82

Simões MF, Lopes DF (1987) Pesquisas arqueológicas no baixo/médio Rio Madeira (Amazonas). Rev Arqueol 4(1):117–134

Thomas RH (1922) The Indian tribes of the northern oriental region of Peru. Manuscript on file, Pitt-Rivers Museum, Oxford

Vázquez F (1981) Relación de la jornada de Pedro de Orsua a Omagua y al Dorado. In: Lope de Aguirre C (1559–1561), González EM, Tur NE (eds) Ediciones de la Universidad de Barcelona, Editorial 7 1/2. Universidad de Barcelona, Barcelona, pp 203–271

Weber RL (1975) Caimito: an analysis of the Late Prehistoric culture of the central Ucayali, eastern Peru. University Microfilms, Ann Arbor

Woods WI, McCann JM (1999) The anthropogenic origin and persistence of Amazonian dark earths. In: Caviedes C (ed) Proc Conf Latin Americanist Geographers. Yearb Conf Latin Am Geogr 25:7–14

Zuñiga de G (1981) Relación muy verdadera de todo lo sucedido en el Rio del Marañon, en la Provincia del Dorado, hecha por el Gobernador Pedro de Orsua. In: Lope de Aguirre C (1559–1561), González EM, Tur NE (eds) Ediciones de la Universidad de Barcelona, Editorial 7 1/2. Universidad de Barcelona, Barcelona, pp 3–29

7 Organic Matter in Archaeological Black Earths and Yellow Latosol in the Caxiuanã, Amazonia, Brazil

MARIA DE LOURDES PINHEIRO RUIVO[1], EWERTON DA SILVA CUNHA[2], and DIRSE CLARA KERN[1]

7.1 Introduction

The National Reserve of Caxiuanã, located in the Lower Amazon region, is a place where environmental preservation is still possible even though its surroundings are densely inhabited. The Ferreira Penna Scientific Station (ECFPn) is a building in the reserve and belongs to the Museu Paraense Emílio Goeldi (MPEG), supporting multidisciplinary research in terms of species, communities, and ecosystems.

The environmental conservation, low demography, and high biodiversity are favorable to support scientific studies. Investigations developed in Caxiuanã are mainly focused on plant and animal biodiversity, due to the dense tropical rainforest which covers a large extent of the scientific station. Most of these studies were published in the book *Caxiuanã*, a publication with 30 works distributed in 5 chapters, with subjects related to the history, human activity, fauna, flora, and geology of Caxiuanã (Lisboa and Ferraz 1999).

Most soils in Caxiuanã are latosols, but there are also gleysoils, planosols, podzols, and archeological black earth (ABE), locally termed *terra preta do índio* (Indian Black Earth). ABE shows characteristics of former human occupation, due to ceramic and lithic artifacts in deeper zones of its profile. This soil has a dark color and high contents of Ca, Mg, Zn, Mn, P, and C (Kern 1996). It also contains a higher quantity of humic material due to the better quality of its organic matter. ABE occurs among all soil types of the Amazon region, but most commonly above latosols, podsols, plintosols, and the so-called *terra roxa estruturada* (Kern 1988). There are a few areas with ABE in Caxiuanã occurring in former orchard sites, with cultivation remains. These areas were studied by Kern (1996), and include the "Manduquinha", "Ponta Alegre", and "Mina II" sites.

Latosols are under the large extent of tropical rainforest and also under some small areas of secondary vegetation, locally termed *capoeira*. Ruivo et al. (2001) studied latosols in Caxiuanã present in sites from the ESECAFLOR (The Impact of Drought on Water and Carbon Dioxide Fluxes from Brazilian

[1] Museu Paraense Emílio Goeldi, Av. Perimetral n. 1901, CEP 66077-530, Campus de Pesquisa, Belém, Pará, Brasil
[2] Instituto de Pesquisa Ambiental da Amazônia, Av. Nazaré, 669 Belém, Pará, Brasil

Rain Forest) experiment, a subproject of the Large-Scale Biosphere--Atmosphere (LBA) Program. The authors described these soils as containing A, B, and C horizons, having a medium drainage and sandy texture. Latosols surrounding the ECFPn, in an area where there is an observation tower for the LBA project have a clayey texture and are poorly drained soils (Cunha 2001). The objective of this work is to study variations in mineral and organic components of soils from five sites located in the ECFPn area and surrounding areas, four being yellow latosols and one ABE.

7.2
Materials and Methods

The study was carried out at the National Reserve of Caxiuanã (central coordinates 1042'30"S and 51031'45"W), where the five sites were chosen. Three sites belong to areas inside the Ferreira Penna Scientific Station, in forest ecosystems, one site is the area where there is a tower for collecting climatic data for the LBA project (TOW site), and two sites are the experimental plots designated to the ESECAFLOR experiment (PA and PB sites) comprising 1 ha. One plot (PA) is designed to suffer the absence of rainwater and the other (PB) is used as a control. Outside the ECFPn, two sites were chosen. One is in an approximately 20-year-old stand of secondary vegetation (*capoeira*) near the Curuá River (CAP site), an area where there is an abandoned heliport pertaining to the ECFPn; and the other site is near Caxiuanã Bay is in the former orchard site where the ABE is located (ABE site). This site belongs to the Brazilian Institute for Environmental Monitoring (IBAMA) and is locally named the "*Manduquinha*" site.

Three trenches were made in each site. Soil samples were collected at depths of 0–5, 5–10, 10–20, and 20–50 cm. The samples were air-dried and sieved to <2 mm. Texture and chemical properties were determined by methods of EMBRAPA (1979). Ca^{2+} and Mg^{2+} were determined by spectrophotometry after extraction with 1 M potassium chloride, organic carbon (OC) was determined with a modified Walkley-Black method, N with the Kjëldahl method, pH in water, acidity by volumetric neutralization after extraction with 1 M calcium acetate, Al^{3+} by volumetric neutralization after extraction with 1 M potassium chloride, and exchangeable P was determined colorimetrically after Mehlich extraction. Chemical properties were submitted to the Tukey test at 5% level and correlation of Pearson. Soil samples were observed with scanning electron microscopy for their micromineralogy.

7.3
Results

Table 7.1 shows the concentrations of OC and total nitrogen and C/N ratios from soils of the studied sites. At the 0–5 cm depth the highest OC concentration was found in the ABE site, differing significantly (P <0.05) from all

Table 7.1. Concentrations (in $g\,kg^{-1}$) of total organic carbon (*OC*), total nitrogen (*N*), and C/N ratios from Caxiuanã soils. Same *letters a, b,* and *c* in columns indicate non-significant differences by the Tukey test at 5% level

Depth (cm)	Site	OC	N	C/N
0–5	ABE	48.53 a	3.17 a	15.73 a
	CAP	19.08 b	1.70 b	11.41 ab
	PA	10.33 c	1.50 b	6.95 b
	PB	13.70 c	1.30 b	11.07 ab
	TOW	26.52 b	2.83 a	9.38 ab
5–10	ABE	42.50 a	2.70 a	19.21 a
	CAP	32.43 ab	1.80 a	12.54 a
	PA	10.52 c	1.67 a	6.39 a
	PB	7.13 c	1.27 a	5.62 a
	TOW	16.93 bc	2.00 a	8.47 a
10–20	ABE	35.80 a	1.87 a	27.64 a
	CAP	14.67 b	1.57 a	9.35 a
	PA	6.60 bc	1.27 a	8.60 a
	PB	4.73 c	1.10 a	4.27 a
	TOW	13.37 b	1.20 a	15.01 a
20–50	ABE	13.40 a	1.23 a	10.79 a
	CAP	6.99 b	1.13 a	6.13 a
	PA	4.35 b	1.03 a	4.21 a
	PB	3.48 b	1.00 a	3.51 a
	TOW	7.35 b	1.00 a	9.74 a

other sites. The CAP and TOW sites had high OC contents as well and do not differ amongst themselves. The PA and PB sites showed the lowest OC concentrations of OC and these did not differ significantly with soil depth. At the CAP site, OC contents increased from the 0–5 cm to the 5–10 cm depth increment. The ABE site contained high amounts of OC down to 10–20 cm depth. At the 20–50 cm depth a medium value of carbon ($13.40\,g\,kg^{-1}$) was found.

The highest nitrogen contents were found in the ABE and TOW sites at 0–5 cm soil depth (Table 7.1). Deeper in the soil, no significant differences could be observed among the various sites, although the concentrations of nitrogen tended to be higher in the ABE site. The highest C/N ratio was observed in the ABE site, although the differences to that of the other sites were statistically not significant at the $P < 0.05$ level.

In Table 7.2 concentrations of exchangeable calcium, magnesium, and phosphorus of the soils from the studied sites are presented. In general, the values of the three elements decreased with depth. High values of calcium, magnesium, and phosphorus were found at all depths in the ABE site, significantly higher than in all other sites. Among all the other sites, showing representative values for latosols in the Amazon region, no significant differences could be observed. The content of calcium was high at the 0–5 cm depth of

Table 7.2. Concentrations (in $g\,kg^{-1}$) of exchangeable calcium (Ca^{2+}), magnesium (Mg^{2+}), and phosphorus (P) from Caxiuanã soils. Same *letters a, b,* and *c* in columns indicate non-significant differences by the Tukey test at 5% level

Depths(cm)	Site	Ca^{2+}	Mg^{2+}	P
0–5	ABE	2324.71 a	352.39 a	41.61a
	CAP	115.65 b	73.26 b	6.54 b
	PA	70.11 b	55.90 b	5.13 b
	PB	94.89 b	106.02 b	4.65 b
	TOW	73.88 b	49.91 b	4.77 b
5–10	ABE	2101.49 a	274.46 a	30.48 a
	CAP	65.74 b	34.82 b	4.77 b
	PA	56.14 b	31.00 b	3.00 b
	PB	56.32 b	35.55 b	2.61 b
	TOW	56.50 b	29.04 b	2.46 b
10–20	ABE	2006.16 a	242.83 a	28.20 a
	CAP	49.96 b	23.77 b	3.27 b
	PA	48.11 b	25.32 b	2.07 b
	PB	52.07 b	25.01 b	1.83 b
	TOW	58.88 b	26.14 b	1.47 b
20–50	ABE	1269.08 a	171.95 a	14.73 a
	CAP	50.91 b	21.18 b	1.11 b
	PA	48.18 b	22.63 b	1.26 b
	PB	49.94 b	20.77 b	0.72 b
	TOW	50.97 b	26.27 b	0.84 b

the CAP site, while the magnesium concentration was high for the PB site at the same depth.

Table 7.3 shows the values of pH, acidity, and exchangeable aluminum of the studied soils. The ABE site had higher pH values (minimum of 5.68 and maximum of 6.13), differing significantly ($P<0.05$) from the other sites at all depths. The latosols (CAP, PA, PB, and TOW sites) were acidic with pH values varying from 3.89–4.56. Significant differences amongst them were found at 20–50 cm depth. The CAP and TOW sites differed significantly amongst themselves, but did not differ from the PA and PB sites. Acidity was higher in the CAP site at 10–20 cm depth. The acidity was also high in the TOW site at 0–5 and 5–10 cm depths. Lower values of aluminum were found in the ABE site and they increased in depth. The CAP and TOW sites do not present significant differences amongst themselves. There were no significant differences at 20–50 cm depth.

Table 7.4 presents the textural composition of the studied soils. The sites ABE, CAP, PA, and PB have high contents of sand with higher contents of coarse sand than fine sand. The ABE site has the lowest content of fine sand. The TOW site has the highest content of clay and is classified as a clay texture soil. In general, latosols presented lower values of silt/clay ratios than the ABE, being deeply weathered soils.

Table 7.3. Values (in cmol kg^{-1}) of pH, acidity ($H^{+}+Al^{3+}$), and exchangeable aluminum (Al^{3+}) from Caxiuanã soils. Equal *letters a, b,* and *c* in columns indicate non-significant differences by the Tukey test at 5% level

Depth (cm)	Site	pH	$H^{+}+Al^{3+}$	Al^{3+}
0–5	ABE	6.13 a	7.09 c	0.20 c
	CAP	4.09 b	14.19 a	2.60 a
	PA	4.17 b	5.61 c	1.70 b
	PB	4.37 b	8.28 bc	1.73 b
	TOW	3.89 b	13.56 ab	2.87 a
5–10	ABE	6.06 a	8.08 bc	0.23 c
	CAP	4.07 b	16.70 a	2.93 a
	PA	4.11 b	4.85 c	1.60 b
	PB	4.17 b	6.86 c	1.50 b
	TOW	3.91 b	11.98 b	2.57 a
10–20	ABE	5.98 a	9.67 a	0.23 c
	CAP	4.32 b	12.57 a	2.73 a
	PA	4.01 b	4.16 c	1.63 b
	PB	4.18 b	5.31 bc	1.53 b
	TOW	4.02 b	9.28 ab	2.07 b
20–50	ABE	5.68 a	8.28 a	1.17 a
	CAP	4.56 b	7.13 ab	2.13 a
	PA	4.16 bc	3.96 c	1.30 a
	PB	4.31 bc	4.62 bc	1.43 a
	TOW	4.03 c	7.26 ab	1.77 a

Table 7.4. Textural composition (in $g\,kg^{-1}$) of the study soils from the Caxiuanã site

Site	Depth (cm)	Fine sand	Coarse sand	Silt	Clay	Silt/clay
ABE	0–5	52.1	539.3	205.0	192.0	1.07
	5–10	54.3	498.2	241.9	205.7	1.18
	10–20	59.5	453.6	240.9	244.9	0.98
	20–50	61.7	403.8	247.9	286.6	0.86
CAP	0–5	15.5	626.2	82.7	134.6	0.61
	5–10	21.2	518.9	96.9	169.9	0.57
	10–20	192.0	435.6	148.5	224.0	0.66
	20–50	159.3	298.3	139.6	402.7	0.35
PA	0–5	251.7	565.3	63.3	119.7	0.53
	5–10	233.7	514.7	99.3	152.4	0.65
	10–20	265.0	489.7	78.3	167.0	0.47
	20–50	281.9	438.1	87.6	192.5	0.46
PB	0–5	257.3	537.9	56.5	148.4	0.38
	5–10	286.4	442.0	82.8	188.8	0.43
	10–20	271.7	474.1	77.0	177.2	0.43
	20–50	296.6	486.3	67.1	150.0	0.45
TOW	0–5	186.1	258.4	139.6	415.9	0.34
	5–10	182.7	241.3	147.7	428.4	0.34
	10–20	170.5	192.7	156.3	480.5	0.32
	20–50	136.9	144.6	107.8	610.6	0.18

Figures 7.1–7.3 show scanning electron microscope images of the investigated soils. The mineralogy was similar for all soils, consisting predominantly of kaolinite in the clay fraction and quartz in the sand fraction, and showed a connection between macropores and organic matter. The TOW soil has many microaggregates and moderate porosity. The microaggregates of the ABE soil are dominated by a high porosity, principally due to the presence of organic matter and the sandy texture.

Organic carbon gave positive correlations with pH, N, Ca^{2+}, Mg^{2+}, P, and the content of coarse sand in the ABE site (Table 7.5). Thus, the content of organic carbon increases in the soil as long as there is an increase in those characteristics. Highly significant negative correlations were obtained between OC and Al^{3+}, fine sand, silt, and clay. In the CAP site, positive-significant correlations were found between organic carbon and acidity ($H^+ + Al^{3+}$), Al^{3+}, N, C/N, Ca, Mg, P, and also the contents of coarse and fine sand (Table 7.6). In the PA, PB, and TOW sites, OC was significantly positively correlated with all chemical characteristics, except for pH, where a non-significant relation was observed. In the PA and PB sites (Tables 7.7, 7.8, and 7.9), OC was significantly negatively correlated with all characteristics, except for the coarse sand content where a significantly positive correlation was observed. In the TOW site, OC contents gave significant positive correlations with the contents of coarse and fine sands, significant negative correlations with clay content, and the correlation with silt showed no significance at the $P < 0.05$ level.

The positive correlations between OC, acidity, and pH are due to the fact that the soil organic matter is the dominant source of acidity in highly weathered tropical soils. However, in the ABE site, the correlations between OC and acidity were non-significant and negative with aluminum. This means that the functional groups of the organic matter in the ABE site are saturated with basic cations instead of protons and Al^{3+} as for the other investigated sites.

Nitrogen showed significant positive correlations with pH, OC, Ca^{2+}, Mg^{2+}, and P in the ABE site, so if pH and these elements increase, there is also a nitrogen increase. The positive correlation between two elements indicates a common source of these elements.

In the CAP site significantly positive correlations of acidity with Al^{3+}, OC, and exchangeable phosphorus were found. Except for pH, in the PA, PB, and TOW sites all chemical characteristics were positively correlated to nitrogen. Aluminum gave significantly negative correlation to pH, N, and the C/N ratio in the ABE site (Table 7.5). This relationship with pH indicates that there is an increase of aluminum content when pH values decrease. Phosphorus correlated positively with pH, C, N, Ca^{2+}, and Mg^{2+} in the ABE site (Table 7.5). Calcium and magnesium in the ABE site presented significantly positive correlations with pH, C, N, and P at the $P < 0.01$ level. Looking at textural characteristics, calcium and magnesium correlated positively with the coarse sand content.

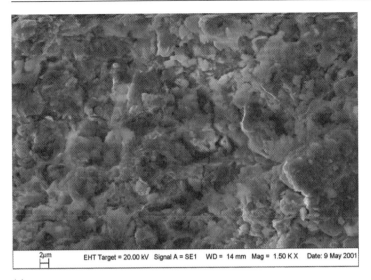

| 2µm | EHT Target = 20.00 kV   Signal A = SE1   WD = 14 mm   Mag = 1.50 K X   Date: 9 May 2001 |

(a)

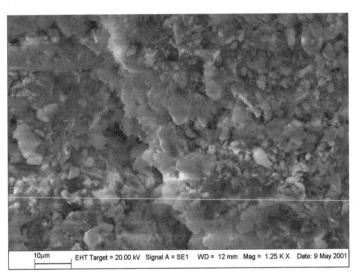

| 10µm | EHT Target = 20.00 kV   Signal A = SE1   WD = 12 mm   Mag = 1.25 K X   Date: 9 May 2001 |

(b)

Fig. 7.1. Scanning electron microscope image of the ABE soil,
showing microaggregation, good porosity, and high organic matter content:
a topsoil; **b** 50 cm depth

(a)

2μm   EHT Target = 20.00 kV   Signal A = SE1   WD = 10 mm   Mag = 2.00 K X   Date: 11 May 2001

(b)

Fig. 7.2. Scanning electron microscope image of the PA=PB soil, showing macroaggregation, less porosity, and less organic matter compared to the ABE soil and a dominance of kaolinite: **a** topsoil; **b** 50 cm depth

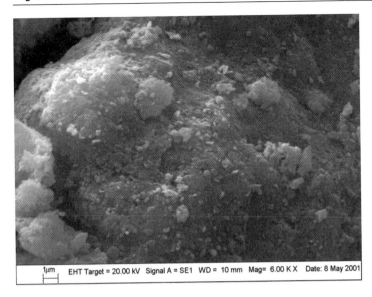

(a)

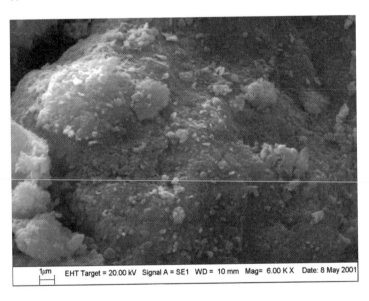

(a)

Fig. 7.3. Scanning electron microscope image of the TOW soil,
showing microaggregation, moderate porosity, and moderate kaolinite content:
a topsoil; **b** 50 cm depth

Table 7.5. Correlation coefficients of chemical and textural characteristics in ABE site. $H^+ + Al^3$ Hydrogen + aluminum; Al^{3+} exchangeable aluminum; OC organic carbon; N nitrogen; C/N carbon/nitrogen ratio; Ca^{2+} exchangeable calcium; Mg^{2+} exchangeable magnesium; P phosphorus; FS fine sand; CS coarse sand; SIL silt; CL clay; ns non-significant

	pH H$_2$O	H+Al^{3+}	Al^{3+}	OC	N	C/N	Ca^{2+}	Mg^{2+}	P	FS	TS	SIL	CL
pH H$_2$O	1												
H$^+$Al^{3+}	−0.31 ns	1											
Al^{3+}	−0.96**	0.02 ns	1										
OC	0.99**	−0.34 ns	−0.95**	1									
N	0.91**	−0.56*	−0.79**	0.91**	1								
C/N	0.49*	0.53*	−0.67 ns	0.47 ns	0.09 ns	1							
Ca^{2+}	0.99**	−0.27 ns	−0.96**	0.99**	0.87**	0.55*	1						
Mg^{2+}	0.93**	−0.57*	−0.81**	0.94**	0.88**	0.34 ns	0.93**	1					
P	0.96**	−0.47 ns	−0.86**	0.96**	0.85**	0.46 ns	0.96**	0.99**	1				
FS	−0.89**	0.69**	0.73**	−0.91**	−0.97**	−0.08 ns	−0.87**	−0.95**	−0.91**	1			
CS	0.95**	−0.60**	−0.81**	0.96**	0.94**	0.26 ns	0.93**	0.99**	0.97**	−0.98**	1		
SIL	−0.68*	0.68**	0.50*	−0.70**	−0.61**	−0.25 ns	−0.70**	−0.89**	−0.86**	0.78**	−0.83**	1	
CL	−0.97**	0.52*	0.86**	−0.97**	−0.98**	−0.26 ns	−0.95**	−0.95**	−0.95**	0.98**	−0.99**	0.72**	1

* and ** Significant at 5 and 1% levels, respectively.

Table 7.6. Correlation coefficients of chemical and textural characteristics in CAP site. $H^+ + Al^3$ Hydrogen + aluminum; Al^{3+} exchangeable aluminum; OC organic carbon; N nitrogen; C/N carbon/nitrogen ratio; Ca^{2+} exchangeable calcium; Mg^{2+} exchangeable magnesium; P phosphorus; FS fine sand; CS coarse sand; SIL silt; CL clay; ns non-significant

	pH H$_2$O	H+Al^{3+}	Al^{3+}	OC	N	C/N	Ca^{2+}	Mg^{2+}	P	FS	TS	SIL	CL
pH H$_2$O	1												
H+Al^{3+}	-0.96**	1											
Al^{3+}	-0.84**	0.95**	1										
OC	-0.98**	0.99**	0.91**	1									
N	-0.96**	0.99**	0.95**	0.97**	1								
C/N	-0.99**	0.99**	0.89**	0.99**	0.96**	1							
Ca^{2+}	-0.65**	0.43 ns	0.16 ns	0.50*	0.46 ns	0.53*	1						
Mg^{2+}	-0.68**	0.46 ns	0.20 ns	0.54*	0.51*	0.56*	0.99**	1					
P	-0.94**	0.82**	0.65**	0.86**	0.86**	0.87**	0.86**	0.88**	1				
FS	-0.42 ns	0.64**	0.80**	0.57*	0.58*	0.55*	-0.42 ns	-0.38 ns	0.09 ns	1			
CS	-0.94**	0.83**	0.65**	0.86**	0.86**	0.87**	0.85**	0.87**	0.99**	0.10 ns	1		
SIL	0.83**	-0.67**	-0.40 ns	-0.75**	-0.63**	-0.78**	-0.86**	-0.85**	-0.86**	0.01 ns	-0.85**	1	
CL	0.96**	-0.93**	-0.84**	-0.93**	-0.97**	-0.93**	-0.66**	-0.70**	-0.95**	-0.35 ns	-0.96**	0.72**	1

* and ** Significant at 5 and 1% levels, respectively.

Table 7.7. Correlation coefficients of chemical and textural characteristics in PA site. $H^+ + Al^3$ Hydrogen + aluminum; Al^{3+} exchangeable aluminum; OC organic carbon; N nitrogen; C/N carbon/nitrogen ratio; Ca^{2+} exchangeable calcium; Mg^{2+} exchangeable magnesium; P phosphorus; FS fine sand; CS coarse sand; SIL silt; CL clay; ns non-significant

	pH H₂O	H⁺+Al³⁺	Al³⁺	OC	N	C/N	Ca²⁺	Mg²⁺	P	FS	TS	SIL	CL
pH H₂O	1												
H⁺+Al³⁺	0.45 ns	1											
Al³⁺	−0.29 ns	0.73**	1										
OC	0.19 ns	0.88**	0.80**	1									
N	−0.02 ns	0.77**	0.84**	0.98**	1								
C/N	0.28 ns	0.94**	0.79**	0.99**	0.94**	1							
Ca²⁺	0.55*	0.98**	0.63**	0.77**	0.63**	0.85**	1						
Mg²⁺	0.48 ns	0.95**	0.66**	0.71**	0.58*	0.80**	0.99**	1					
P	0.37 ns	0.99**	0.77**	0.84**	0.74**	0.90**	0.98**	0.98**	1				
FS	−0.03 ns	−0.67**	−0.70**	−0.94**	−0.97**	−0.88**	−0.51*	−0.43 ns	−0.60**	1			
CS	0.17 ns	0.96**	0.90**	0.90**	0.85**	0.94**	0.90**	0.90**	0.97**	−0.72**	1		
SIL	−0.09 ns	−0.48 ns	−0.45 ns	−0.11 ns	−0.02 ns	−0.22 ns	−0.60**	−0.71**	−0.60**	−0.21 ns	−0.53*	1	
CL	−0.23 ns	−0.97**	−0.86**	−0.88**	−0.82**	−0.93**	−0.93**	−0.93**	−0.99**	0.68**	−0.99**	0.56*	1

* and ** Significant at 5 and 1% levels, respectively

Table 7.8. Correlation coefficients of chemical and textural characteristics in PB site. $H^+ + Al^3$ Hydrogen + aluminum; Al^{3+} exchangeable aluminum; OC organic carbon; N nitrogen; C/N carbon/nitrogen ratio; Ca^{2+} exchangeable calcium; Mg^{2+} exchangeable magnesium; P phosphorus; FS fine sand; CS coarse sand; SIL silt; CL clay; ns non-significant

	pH H$_2$O	H$^+$+Al^{3+}	Al^{3+}	OC	N	C/N	Ca^{2+}	Mg^{2+}	P	FS	CS	SIL	CL
pH H$_2$O	1												
H$^+$+Al^{3+}	0.22 ns	1											
Al^{3+}	0.39 ns	0.87**	1										
OC	0.43 ns	0.96**	0.95**	1									
N	−0.11 ns	0.93**	0.67**	0.80**	1								
C/N	0.48 ns	0.94**	0.96**	0.99**	0.75**	1							
Ca^{2+}	0.58*	0.89**	0.96**	0.98**	0.66**	0.99**	1						
Mg^{2+}	0.56*	0.90**	0.96**	0.98**	0.68**	0.99**	0.99**	1					
P	0.26 ns	0.98**	0.95**	0.98**	0.86**	0.97**	0.93**	0.94**	1				
FS	−0.16 ns	−0.73**	−0.94**	−0.80**	−0.56*	−0.82**	−0.81**	−0.81**	−0.85**	1			
CS	0.83**	0.46 ns	0.76**	0.58**	0.10 ns	0.73**	0.82**	0.80**	0.58*	−0.66**	1		
SIL	−0.94**	−0.38 ns	−0.64**	−0.61**	−0.02 ns	−0.66**	−0.76**	−0.74**	−0.47 ns	0.49*	−0.97**	1	
CL	−0.97**	−0.07 ns	−0.34 ns	−0.31 ns	0.29 ns	−0.37 ns	−0.49*	−0.47 ns	−0.15 ns	0.18 ns	−0.86**	0.94**	1

$\star$ and $\star\star$ Significant at 5 and 1% levels, respectively.

Table 7.9. Correlation coefficients of chemical and textural characteristics in TOW site. $H^{+}+Al^{3}$ Hydrogen + aluminum; Al^{3+} exchangeable aluminum; OC organic carbon; N nitrogen; C/N carbon/nitrogen ratio; Ca^{2+} exchangeable calcium; Mg^{2+} exchangeable magnesium; P phosphorus; FS fine sand; CS coarse sand; SIL silt; CL clay; ns non-significant

	pH H$_2$O	H$^+$+Al^{3+}	Al^{3+}	C	N	C/N	Ca^{2+}	Mg^{2+}	P	AF	AG	SIL	ARG
pH H$_2$O	1												
H$^+$+Al^{3+}	−0.96**	1											
Al^{3+}	−0.97**	0.99**	1										
C	−0.86**	0.96**	0.96**	1									
N	−0.90**	0.96**	0.96**	0.99**	1								
C/N	−0.63**	0.82**	0.79**	0.84**	0.76**	1							
Ca^{2+}	−0.66**	0.82**	0.81**	0.95**	0.92**	0.80**	1						
Mg^{2+}	−0.75**	0.83**	0.85**	0.95**	0.96**	0.68**	0.97**	1					
P	−0.86**	0.85**	0.93**	0.99**	0.99**	0.74**	0.94**	0.98**	1				
AF	−0.83**	0.93**	0.91**	0.86**	0.82*	0.94**	0.72**	0.67**	0.78**	1			
AG	−0.94**	0.99**	0.99**	0.93**	0.92**	0.85**	0.78**	0.78**	0.88**	0.96**	1		
SIL	−0.36 ns	0.53*	0.50 ns	0.47 ns	0.37 ns	0.86**	0.39 ns	0.22 ns	0.32 ns	0.81**	0.63**	1	
ARG	0.84**	−0.93**	−0.92**	−0.87**	−0.83**	−0.93**	−0.72*	−0.67**	−0.78**	−1.00	−0.97**	−0.80**	1

* and ** Significant at 5 and 1% levels, respectively.

7.4
Discussion

The high OC contents in the ABE site (Table 7.1) are in agreement with data found by other investigations of ABE soils (e.g. Kern 1988, 1996). This indicates that the main *input* of organic matter in these soils is of anthropogenic origin. da Costa and Kern (1999) reported that the black earth at Caxiuanã developed over yellow latosols and was formed from weathering decomposition of lateritic ferro-aluminous crust. This crust originally derived from Cretaceous sediments (Alter do Chão Formation, Amazon Sedimentary Basin). However, it is necessary to consider the *input* originating from the vegetation as an important factor maintaining the stability of this ecosystem. Soils of the ESECAFLOR experiment (the PA and PB sites) have lower values of OC than the CAP and TOW sites. The values of OC decreased gradually at depth at the PA and PB sites, while there is an accumulation of OC in the TOW site at 0–5 cm depth, and a sudden decrease in carbon from the 0–5 to the 5–10 cm depth. Ruivo et al. (2001) found a similar situation when studying the same area. The authors attributed this to the drainage variation of these soils. Martins and Cerri (1986) also found an interference in the carbon distribution along the soil profile related to the drainage conditions in soils of a forest ecosystem in the Amazon region. Better distribution of OC along the soil profile in the PA and PB sites also indicates better conditions of carbon cycling in the ecosystem. The high nitrogen contents at the TOW site at 0–5 cm depth were also associated with accumulation due to poor drainage. The low C/N ratios observed in the PA, PB, and TOW sites are typical for soils from native ecosystems, where the proportion between C and N is almost constant. It also indicates an equilibrium state (Kern 1988). The amounts of exchangeable phosphorus (Table 7.2) showed that this element decreased in depth in all sites, being highest in the ABE site, in accordance with Costa and Kern (1999), who studied the geochemistry of ABE from Caxiuanã. The latosols did not present statistical differences in the contents of phosphorus; however, medium values were found in the CAP site at 0–5 cm and lower values in the following depths. The sites PA, PB, and TOW have low values of this element in all depths. Phosphorus, like nitrogen, originates from the organic matter, which is why higher values are generally found at the A_1 horizon and its transport to the B horizon is possible (Britez et al. 1997).

Amazonian soils, mainly latosols, are deeply weathered. Elements such as Ca^{2+} and Mg^{2+} are therefore leached. The sites CAP, PA, PB, and TOW have lower values of these elements than the ABE site. The highest contents of calcium and magnesium in the ABE are pointed out as products of deposition from the native population. These materials deposited in the soil were probably rich in calcium and magnesium (e.g. bones, shells, and other disposed materials of animal and vegetal origin). Large amounts of non-consumed organic matter accumulated in the soil have remained in place, increasing the contents of calcium, magnesium, and phosphorus in the soil (Kern 1996).

Stimulation of microbial activity by these *inputs* may lead to accelerated mineralization of substances less stable to degradation (such as polysaccharides and proteins), and thus to a relative enrichment of humic fractions rich in aromatic constituents (Zech et al. 1990). The data related to the soil acidity (Table 7.3) show that the soil from the ABE site has a moderate acidity, while latosols (CAP, PA, PB, and TOW sites) are extremely acid. The $H^+ + Al^{3+}$ predominance in the exchange complex indicates a low base saturation, and deficiencies of the elements Ca^{2+}, Mg^{2+}, and K^+ may occur. The high pH values in the ABE site are due to the higher contents of calcium and magnesium, and also indicate a better quality of the soil organic matter in the ABE. The lowest values of aluminum indicate a low degree of toxicity of this element, different from the other sites which showed higher values. According to the chemical properties, decreasing fertility of the investigated soils is in the order ABE>-CAP≅TOR>PA=PB. The textural composition presented in Table 7.4 indicates that the soils from the ABE and CAP sites have a loamy texture, while PA and PB have a sandy texture. The TOW site has a clay texture, which makes the drainage poor through the soil profile, promoting the accumulation of some elements, as was previously shown. The low content of silt is a common characteristic of deeply weathered soils (Klamt and Van Reeuwijk 2000). The highest values of the silt/clay indicate that the ABE is more recent than latosols, due to the higher silt contents. The latosols presented characteristics of deeply weathered soils, common of Amazonian soils.

The ABE site has significantly higher contents of OC, N, P, Ca^{2+}, and Mg^{2+} and low conditions of acidity, different from the yellow latosols, which presented characteristics of weathered soils. Latosols also showed equilibrium conditions from undisturbed ecosystems, especially the PA and PB sites, which presented uniform features, favorable to evaluate the conditions of nutrient cycles before the introduction of the ESECAFLOR experiment. In the TOW site the bad drainage conditions promoted an accumulation of organic matter expressed in high OC and N contents at 0–5 cm depth.

The highly positive significant correlations between pH and OC, N, Ca^{2+}, Mg^{2+}, and P (Table 7.5) in the ABE show that the increase of these elements in soil contributes to the increase of pH values, mainly Ca^{2+} and Mg^{2+}, due to the dominance of these elements in soil exchange complex at these sites. In contrast, in the latosols, increasing amounts of OC were positively correlated to soil acidity, indicating that soil organic matter is the dominant contributor to soil acidity in highly weathered soils of the humid tropics. In the case of nitrogen, there is a relation mentioned by Killham (1994), where higher plants assimilate nitrogen as nitrates (by *nitrate-reductase* activity), releasing bicarbonate and/or hydroxyl ions in order to keep the charge balance, thus affecting pH values by increasing them. On the other hand, increases of pH values result in higher nitrogen availability.

The correlation between phosphorus and organic carbon in the ABE site indicates that the organic matter may influence phosphorus availability. Actually, soil organic matter also contributes to the availability of other elements. Pabst (1991) studied ABE in Belterra (Pará) and found significantly positive correlations of soil organic matter and calcium, magnesium, and potassium, reflecting the great importance of soil organic matter for complexing exchangeable cations.

References

Britez RM, Santos Filho A, Reissmann CB, Silva SM, Athayde SF, Lima RX, Quadros RMB (1997) Nutrientes no solo de duas florestas na planície litorânea Rev Bras Ciênc Solo 21:625–634

Costa da ML, Kern DC (1999) Geochemical signatures of tropical soils with archaeological black earth in the Amazon, Brazil. J Geochem Explor 66:369–385

Cunha da Silva E (2001) Estudo comparativo entre *Terra Preta* Arqueológica e Latossolos Amarelos através da variação dos componentes orgânicos e minerais na região de Caxiuanã (Melgaço, PA). Master's Diss, Faculdade de Ciências Agrárias do Pará, Belém

EMBRAPA (1979) Manual de métodos de análises de solo. Serviço Nacional de Levantamento e Conservação de Solo, Rio de Janeiro

Kern DC (1988) Caracterização pedológica de solos de *Terra Preta* Arqueológica na região de Oriximiná- Pará. Master's Diss, Faculdade de Agronomia da Universidade Federal do Rio Grande do Sul, Porto-Alegre, 231 pp

Kern DC (1996) Geoquímica e pedogeoquímica em sítios arqueológicos com *Terra Preta* na Floresta Nacional de Caxiuanã (Portel- PA). PhD Thesis, Universidade Federal do Pará/Centro de Geociências/Curso de Pós-graduação em Geologia e Geoquímica, Belém

Killham K (1994) The ecology of soil nutrient cycling. Soil ecology. Cambridge University Press, Cambridge, 242 pp

Klamt E, Van Reeuwijk LP (2000) Evaluation of morphological, physical and chemical characteristics of Ferralsols and related soils. Rev Bras Ciênc Solo 24:573–587

Lisboa PLB, Ferraz M das G (1999) Estação Científica Ferreira Penna: Ciência e desenvolvimento sustentável na Amazônia. Museu Paraense Emílio Goeldi, Belém, 151 pp

Martins da Silva PF, Cerri CC (1986) O solo de um ecossistema natural de floresta localizado na Amazônia Oriental. I. Caracterização física e química. In: Proc Symp Trópico Úmido, I, Belém, Embrapa-CPATU, Anais, 6v. Embrapa-CPATU, Documentos 36, pp 271–286

Pabst E (1991) Critérios de distinção entre *Terra Preta* e Latossolo na Região de Belterra e os seus significados para discussão pedogenética. Bol Mus Par Emílio Goeldi Ser Antropol 7:5–19

Ruivo MLP, Pereira SB, Bussetti EPC, Costa RF, Quanz B, Nagaishi TY, Meir P, Mahli Y, Costa AL (2001) Variações no solo e no fluxo de CO_2 nos sítios do ESECAFLOR, Caxiuanã, PA. In: Proc Symp Geologia da Amazônia, UFPA/SBG, Belém, CD-ROM dos Resumos Expandidos

Zech W, Haumaier L, Hempfling R (1990) Ecological aspects of soil organic matter in tropical land use. In: McCarthy P, Malcolm RL, Bloom PR (eds) Humic substances in soil and crop sciences: selected readings. American Society of Agronomy and Soil Science Society of America, Madison, Wisconsin, pp 187–202

8 Sequential P Fractionation of Relict Anthropogenic Dark Earths of Amazonia

Johannes Lehmann[1], Carla Vabose Campos[2],
Jeferson Luiz Vasconselos de Macêdo[2], and Laura German[3]

8.1
Introduction

Dark earths which are rich in small artifacts can be found in a wide range of environmental settings in the Amazon Basin. While various conflicting theories have been proposed about the origin of these soils (Smith 1980), it is now widely accepted that they formed under anthropogenic influence (Sombroek 1966; Smith 1980; Woods et al. 2000). These relict Anthrosols are often the only testimony of pre-Columbian settlements and can provide important information about the former inhabitants (Vacher et al. 1998). Amazonian Dark Earths have high soil organic matter and nutrient contents such as P (Sombroek 1966; Smith 1980; Kern and Kämpf 1989), and are therefore highly fertile. Phosphorus availability is the most important constraint to crop production in central Amazonia (Lehmann et al. 2001a) and high P contents are the primary reason for the high production potential of these Anthrosols. Farmers value these soils in many areas of Amazonia, and some authors claim that continuous cropping is possible for 40 years or more without fallowing (Petersen et al. 2001). Whether the so-called *terra preta de índio* soils were intentionally created for agricultural purposes or whether they formed as a result of habitation is still under debate (McCann et al. 2001). In order to understand the origin of these soils, it is important to know what type of organic input caused the high organic matter and P contents, and under which conditions. More information is needed about the properties of these soils to answer such questions. This would provide clues about the livelihood of the former inhabitants as well as strategies for future soil use.

Readily extractable P contents have been widely used for the identification and study of anthropogenic soils (Arrhenius 1931; Eidt 1977; Woods 1977; Wells et al. 2000). The spatial extent of anthropogenic activities can be determined without large excavations and provide important information even in the absence of artifacts. A sequential fractionation scheme that distinguishes between different soil P pools has been applied to archaeological sites, yield-

[1] Department of Crop and Soil Sciences, College of Agriculture and Life Sciences, Cornell University, Ithaca, New York 14853, USA
[2] Embrapa Amazônia Ocidental, C.P. 319, 69011-970 Manaus, Brasil
[3] World Agroforestry Centre (ICRAF), P.O. Box 26416 (ICRAF/AHI), Kampala, Uganda

ing more reliable information about settlement history than total or readily available P contents (Eidt 1977; Woods 1977; Lillios 1992). Increasing use of a fractionation procedure developed by Hedley et al. (1982) and modified by Tiessen and Moir (1993) for use in agricultural research has prompted the present study. The procedure addresses the need to explain the high fertility of *terra preta* soils from the viewpoint of current soil use as well as the origins of P deposited in these soils. The Hedley fractionation technique distinguishes between several inorganic and organic soil P pools of increasing recalcitrance and has frequently been used to quantify biochemical cycling of P in agricultural (Friesen et al. 1997) and natural (Cross and Schlesinger 1995) ecosystems. This fractionation procedure was developed for mineral soils and has rarely been applied to soils with high organic matter contents. Since the relict Anthrosols additionally have very high P contents (Sombroek 1966; Kern and Kämpf 1989), the extraction efficiency may be insufficient using the procedure recommended for the Hedley fractionation. In this experiment, we studied the applicability and usefulness of the sequential fractionation modified by Tiessen and Moir (1993) for *terra preta* soils of Amazonia. Different ratios of soil-to-extractant were compared between two Anthrosols and two fertilized Ferralsols from central Amazonia.

8.2
Materials and Methods

8.2.1
Study Sites

This study compares different P pools of two anthropogenic Dark Earths with unfertilized and fertilized *terra firme* soils from central Amazonia. One Anthrosol was collected near Rio Preto da Eva, 60 km north of Manaus, Brazil. The second Anthrosol was obtained from the lower Rio Negro in the community of Marajá. The Ferralsols were sampled at the station of the Embrapa Amazonia Ocidental, 29 km north of Manaus. The average temperature at the Embrapa is 26 °C and average precipitation is 2,503 mm year^{-1} (1971–1993) with a maximum between December and May. The natural vegetation is a tropical rainforest. The Xanthic Ferralsols (FAO 1990) are clayey, strongly aggregated, with medium organic C contents, and low pH values (Table 8.1). The Fimic Anthrosols are sandy loams to sands, with high organic C contents, and moderately acid pH.

Samples with both high and low available P contents were selected from the Ferralsols and the Anthrosols (Table 8.1). Composite soil samples from the two Anthrosols were randomly collected from forest sites. The fields have never received commercial fertilizers. Soils from the fertilized Ferralsols were obtained from a mixed tree crop plantation with *Bixa orellana* L. (urucum; annatto). These samples were collected at two points in 0.5 m distance from two trees (four subsamples). The soil representing low additions of inorganic

Table 8.1. Characterization of a fimic Anthrosol ("*terra preta de índio*") and xanthic Ferralsol from central Amazonia. *n.d.* Not determined

Soils	Sampling depth (m)	pH (H$_2$O)	Sand (%)	Silt (%)	Clay (%)	TOC (g kg^{-1})	P$_{available}$ (mg kg^{-1})
Anthrosol Marajá	0–0.1	5.7	58.7	17.0	24.3	n.d.	25.4[a]
Anthrosol Rio Preto da Eva	0–0.1	5.7	71.0		29.0	84.7	6.5[b]
Ferralsol – high fertilization	0–0.1	4.7	21.4[c]	19.6	59.0	36.0	142.6[b]
Ferralsol – low fertilization	0.1–0.2	4.2	27.2	14.9	57.9	21.2	4.3[b]
Ferralsol – unfertilized	0–0.05	4.1	21.4	19.6	59.0	40.6	4.2[b]

[a] Mehlich-1 extraction (Soil and Plant Analysis Council 1999).
[b] Mehlich-3 extraction (Mehlich 1984).
[c] Particle size distribution determined at a nearby soil pit (J. Marques, unpubl. data).

P was collected at a depth of 0.1–0.2 m from trees receiving an application of 11.9 g P tree^{-1} year^{-1}, whereas the soil representing high additions of inorganic P was obtained from 0–0.1 m depth with an application of 59.4 g P tree^{-1} year^{-1}. Phosphorus was applied as triple super phosphate and split between a December and a May application. Data for unfertilized soils were taken from a replicated assessment of soils under primary forest presented earlier (n=3; Lehmann et al. 2001b). The soils show a wide range of available P contents (Table 8.1).

8.2.2
Soil P Fractionation and Analysis

The soils were air-dried and sieved to 2 mm. Samples were sequentially extracted according to a modified Hedley procedure (Fig. 8.1; Hedley et al. 1982; Tiessen and Moir 1993). The different fractions have been assigned to different soil P pools by various authors (Hedley et al. 1982; Tiessen and Moir 1993; Cross and Schlesinger 1995), but uncertainty about whether these pools can be clearly identified and quantified through sequential extraction prevails. In this publication, we will therefore interpret trends between soils rather than compare different soil P pools. For this reason, Fig. 8.1 gives only the most robust differentiations between pools, which can be unambiguously interpreted.

Four different ratios of soil to extractant were used, 1:1,000, 1:200, 1:100, and 1:40 (10, 50, 100, or 250 mg soil with 10 ml of solution). Tiessen and Moir (1993) recommended 1:60. Two resin strips (each 6×15 mm; Prod 55164 2S, BDH Laboratory Supplies, Poole, UK) were added to 10 ml of distilled water and shaken for 16 h. After retrieving the strips and washing adhering soil with distilled water into the centrifuge tube, P was extracted from the resin with 10 ml 0.5 M HCl for 16 h. The soil suspension was centrifuged at

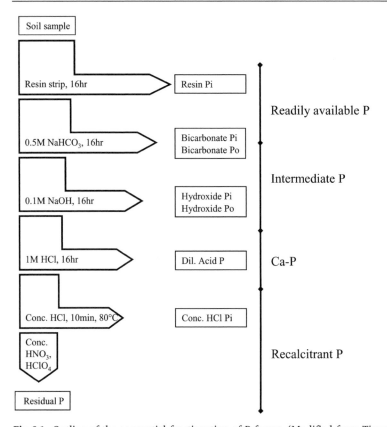

Fig. 8.1. Outline of the sequential fractionation of P forms. (Modified from Tiessen and Moir 1993)

2,000 rpm for 10 min and the supernatant was discarded. Consecutively, extractions were made with 10 ml of 0.5 M NaHCO$_3$ (adjusted to pH 8.5), 0.1 M NaOH, and 1 M HCl, each separately shaken for 16 h and centrifuged at 2,000 rpm (Fig. 8.1). The supernatant was carefully decanted and stored in the refrigerator for analysis. Two milliliters of concentrated HCl was added to the soil and the tubes were placed in a water bath at 80 °C for 10 min. After removal, the suspension was allowed to cool down for 1 h, while shaking the tubes every 15 min by hand. After adding 1 ml of concentrated HCl to the same tubes, they were centrifuged at 2,000 rpm for 10 min and the supernatant was poured into a 10-ml volumetric flask. This was repeated with 2 ml of distilled water. The volumetric flask was then made up to volume with distilled water. After adding 3 ml of concentrated HNO$_3$ and 1 ml of concentrated HClO$_4$ the soil was transferred to a crucible and placed in a sand bath at 200 °C for 16 h. After the soil cooled down, 2 ml of 5 M HNO$_3$ was added to the crucible, filtered into a 10-ml volumetric flask and made up to volume with HNO$_3$.

Inorganic P in the resin-, dilute-acid-, acid-, and residual-P extracts was measured directly using the molybdate ascorbic acid method (Murphy and Riley 1962) on a spectrophotometer at 712 nm. For the alkaline extracts, 1.2 and 0.3 ml of 0.9 M H_2SO_4 was added to 2 ml of bicarbonate and hydroxide extract, respectively. After centrifugation at 4,000 rpm for 20 min and cooling in a refrigerator for 30 min, inorganic P was measured as described above. This procedure separates any organic precipitates from the solution (Tiessen and Moir 1993). Organic P in the bicarbonate and hydroxide extracts was calculated as the difference between total and inorganic P, since organic P cannot be determined directly. Total P was analyzed as inorganic P after digestion with ammonium persulfate and sulfuric acid, followed by autoclavation (Tiessen and Moir 1993). An aliquot of 1 ml bicarbonate or hydroxide extract was digested with 0.1 g ammonium persulfate and 2 ml 0.9 M H_2SO_4 in an autoclave for 2 h. Inorganic P was determined as described above for the resin and acid extracts.

8.3
Results and Discussion

8.3.1
Extraction Efficiency by Sequential Fractionation

The yield of total P and the distribution between pools in the Anthrosol did not change significantly over a wide range of soil-to-extractant ratios, considering the large amounts of P present in this soil (Fig. 8.2). With high ratios, readily extractable P in the resin-P and bicarbonate-Pi fractions of the

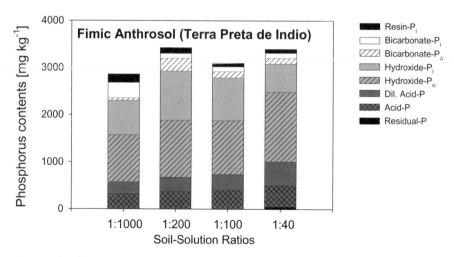

Fig. 8.2. Phosphorus contents in P pools of a Fimic Anthrosol ("*terra preta de indio*"; Amazonian Dark Earth) from central Amazonia with different soil-to-extractant ratios ($n=3$; for 1:40 $n=1$; CV<10%)

Anthrosol may have been overestimated. It can be concluded, however, that the fractionation procedure applied is efficiently extracting the targeted P pools despite the high P and organic matter contents. The proportion of soil to extractant of 1:60 recommended in the standard procedure (Tiessen and Moir 1993) is suitable for the studied Anthrosols. The amount of soil for the extraction should not be less than 0.25 g as the heterogeneity of the soil sample will make it difficult to obtain a representative subsample.

In the Ferralsol, a higher proportion of extractant increased the total P extracted and changed the distribution of P pools. More readily available P (resin- and bicarbonate-P) as well as recalcitrant P (dilute-acid-, acid-, and residual-P) were obtained with these higher ratios. With higher proportions of the extractant, the initial steps of the sequential fractionation yielded more P at the expense of intermediate fractions. Most likely, inorganic P from interlattice layers of clay minerals or from strongly adsorbed P forming bridging ligands with Fe- and Al-oxides may have been extracted at an earlier step in the fractionation. This process was not important for the Anthrosols, which contain very low amounts of clay-sized particles. However, it shows clearly that sequential extraction methods have to be evaluated with caution, as earlier extraction steps have important effects on fractions extracted later in the procedure. Independent analyses of the fractions and parallel extractions are needed to verify the results of sequential fractionations.

8.3.2
Phosphorus Distribution in Soil Pools of Anthropogenic Dark Earths in Central Amazonia

Total P contents of the four soils varied widely between 193 and 3,097 mg kg^{-1} (Table 8.2). These values are far above the ones that are typically found in acid upland soils of central Amazonia, which range from 40–100 mg P kg^{-1} from several different sites (Lehmann et al. 2001a). The Anthrosols must have received large amounts of P during pre-Colombian occupation, whereas the Ferralsols studied here received large and sustained fertilization of 12–59 g P tree^{-1} year^{-1} for 7 years. Glaser (1999) reported total P values of 980–2,170 mg kg^{-1} for five different Anthrosols from central Amazonia.

While increasing total P content, the P added to the Anthrosols by pre-Columbian Indians increased not only readily available P pools but also more stable P pools at least in the Anthrosol with the high total P content. A higher proportion of the inorganic P (triple super phosphate) applied to the Ferralsol remained in readily available forms and was extracted by resin and bicarbonate. An important difference between the pool distribution of the Anthrosol and the fertilized Ferralsol was the higher proportion of hydroxide- and dilute-acid-P in the Anthrosol, as well as the virtual absence of residual P in these soils.

A comparison of the Anthrosol with an unfertilized Ferralsol from a primary forest site (Fig. 8.3) demonstrates this pattern more clearly. The pool

Table 8.2. Inorganic and organic P pools in a fimic Anthrosol ("*terra preta de índio*") and a fertilized xanthic Ferralsol from the central Amazon (only soil-to-extract ratio of 1:100; $n=3$; means ± standard errors; in mg kg^{-1}). *Sum inorganic* Bicarbonate-P_i+hydroxide-P_i+dil. Acid-P+acid-P+residual-P; *Sum organic* bicarbonate-P_o+hydroxide-P_o

Soils	Resin-P	Bicarb-P_i	Bicarb-P_o	Hydr-P_i	Hydr-P_o	Dil. Acid-P	Acid-P	Res-P	Sum inorganic	Sum organic	Total P
Anthrosol Marajá	73.5±4.9	102.4±2.2	131.4±1.7	914.2±12.5	1137.4±29.4	343.4±0.7	394.7±0.8	0.0±0.0	1754.6±13.9	1342.2±29.5	3096.5±13.9
Anthrosol Rio Preto da Eva	6.4±0.4	27.5±1.9	42.5±6.0	74.0±2.6	24.9±2.8	5.5±0.7	11.8±0.7	0.0±0.0	118.8±4.2	73.8±8.1	192.5±7.0
Ferralsol – high fertilization	40.5±2.1	55.0±2.5	144.1±1.3	351.4±2.6	283.7±5.1	29.6±1.7	64.0±2.1	9.3±0.4	509.4±5.0	468.3±6.9	977.7±5.7
Ferralsol – low fertilization	4.5±0.4	12.1±1.0	83.4±2.7	59.3±2.0	47.5±4.2	1.2±0.4	20.4±0.8	7.2±0.4	100.2±1.4	119.5±19.0	219.7±17.7

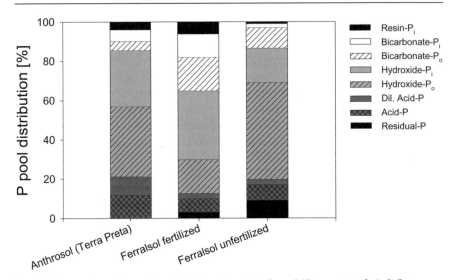

Fig. 8.3. Proportion of P pools to total P in a Fimic Anthrosol ("*terra preta de indio*"; Amazonian Dark Earth) and a fertilized Ferralsol (tree crop plantation – high fertilization) (means of three soil-solution ratios), compared to an unfertilized Ferralsol (mean of forest sites from Lehmann et al. 2001b) in the central Amazon

distribution shifts from plant-available P in the Anthrosol to highly occluded P in the unfertilized Ferralsol. Phosphorus availability seemed to be controlled to a greater extent by intermediate P pools in the Anthrosol and by more recalcitrant P pools in the Ferralsol. This led to a higher proportion of readily plant-available P in the Anthrosol relative to the unfertilized Ferralsol. More recalcitrant P (in the residual-P fraction) can be explained by a higher amount of clay minerals and Fe- and Al-oxides in the Ferralsols, which is confirmed by the particle size distribution (Table 8.1). In support of this explanation, higher proportions of P in coarse particle size fractions such as sand were found in Anthrosols than in Ferralsols (Glaser 1999).

8.3.3
Sources of P in Pre-Colombian Anthrosols

What can the P pool distribution tell us about the types of inputs that led to high levels of P in these Anthrosols? Eidt (1977) and Woods (1977) introduced a sequential fractionation technique that was successfully used to estimate settlement history and soil use. Equal proportions of available P (estimated as hydroxide- and citrate-extractable), occluded P (dithionite-extractable), and Ca-P (HCl-extractable) were shown to be indicative of archaeological soils and intensive human occupation in the past. The modified Hedley fractionation employed in the present study did not yield suitable analoga to the fractions described above. Some patterns of P distribution nevertheless show clear indications of prior occupation: (1) the dominant proportion (65%) and high amounts ($>2,000$ mg kg^{-1}) of P in inorganic and organic hydroxide fractions indicate large inputs of stable organic P compounds; (2) the low importance of residual-P suggests little geogenic or occluded P; and (3) the relatively low proportion of available P (resin- and bicarbonate-Pi) compared to the fertilized Ferralsol, demonstrating the absence of recent inorganic fertilizer inputs.

Judging from the P distribution as well as from the known resources available in the region, it is unlikely that mineral amendments were used by Amerindian inhabitants prior to European contact. More likely is the use of organic P sources as soil amendments, which is supported by the large amount of organic P in these soils. Significant amounts of dilute-acid-P indicate that Ca-phosphates (Ca-P) are present in the Anthrosols. Since pH values of below 6 (Table 8.1) do not promote the formation of Ca-phosphate, P was most likely added in the form of Ca-P. Livestock manure was shown to increase total P as well as available and Ca-P pools (Solomon and Lehmann 2000), but livestock were not present in the Amazon during pre-Columbian times (Gilmore 1963). Human excreta are a possible source of manure with high P contents, and turtle farming has been mentioned as a way of generating manure among early riverine inhabitants (W. Sombroek, pers. comm.). Fish manure in trout farms has been shown to contain high P concentrations of 25.4 g kg^{-1} with N-to-P ratios close to unity, as well as high Ca concentrations of 69.9 g kg^{-1} (Naylor et al. 1999). Based on the current scholarship, it seems unlikely that Amerindian groups alternatively raised fish in ponds.

A more probable source of organic P is from application of kitchen residues containing high proportions of fish. Since these Anthrosols are usually situated on bluffs near rivers (Denevan 1996), fish may have constituted a significant part of the diet. Juvenile tilapia in an aquarium study contained between 20 and 30 g P kg^{-1} (Mbahinzireki et al. 2001). Most of the P (85%) in fish is found in the skeleton (Lall 1991 and Persson 1987 cited in Rønsholdt 1995), which can contain as much as 50 g P kg^{-1} as reported for mackerel (Shimosaka 1999). Fish wastes were shown to contain high P and Ca contents, which were greater in skin and bone fractions (61 and 106 g kg^{-1}) than in total fish processing waste (23 and 45 g kg^{-1}) or in deboned waste (12 and 19 g kg^{-1} for P and Ca, respectively; Rathbone et al. 2001). Thus, inedible parts of fish, such as the bones, had especially high P and Ca concentrations. After cooking the fish, the relative content of P in bones increases even more and is present largely as hydroxo-apatite (Shimosaka 1999). This apatite or similar Ca-phosphates may be the source of Ca-P and the large amounts of P found in *terra preta* soils. Since amounts of P input into villages were as high or higher in fish bones than fish meat, a direct application of fish residues to soil is more probable than through human excreta. Therefore, much of the P in the relict Anthrosols was most likely derived from fish residues and transformed from Ca-P to organic P through microbial activity and simultaneous additions of organic matter.

8.4
Conclusions

The sequential extraction method described in this chapter is suitable for analysis of anthropogenic dark earths of the Amazon, despite the large amounts of P and organic matter in these soils. In comparison to a Ferralsol, the Anthrosols contain much more P which is less abundant in highly recalcitrant pools. However, in comparison to a highly fertilized Ferralsol, the Anthrosols also contained less P in easily extractable and plant-available forms. The dark earths were shown to have the highest fractions of P in intermediate soil P pools. A high amount and proportion of dilute-acid-P in the Anthrosols is indicative of P associated with Ca, suggesting that the high quantities of P could have derived from fish residues. Most of the P in fish is found in the bones, which have high concentrations of P mainly present in the form of Ca-P. The large amounts of P found in these relict Anthrosols from pre-Columbian occupation are likely to result from an application of fish waste. Further studies should include electron microscopy to identify Ca-P compounds in soil. Relative abundance of Ca-P may then be used to examine local diets, market strategies, and long-term turnover of soil P. With sufficient information about settlement history, questions may be addressed about the type of contact between riverine and upland settlements.

Acknowledgements. This study was supported by a fellowship of the Brazilian National Research Agency (Cnpq) to Carla Vabose Campos. The useful comments of Murray McBride and William Woods are gratefully acknowledged.

References

Arrhenius O (1931) Die Bodenanalyse im Dienst der Archäologie. Z Pflanzenernähr Düng Bodenk 10:185–190

Cross AF, Schlesinger WH (1995) A literature review and evaluation of the Hedley fractionation: applications to the biochemical cycle of soil phosphorus in natural ecosystems. Geoderma 64:197–214

Denevan WM (1996) A bluff model of riverine settlement in prehistoric Amazonia. Ann Assoc Am Geogr 86:654–681

Eidt RC (1977) Detection and examination of anthrosols by phosphate analysis. Science 197:1327–1333

FAO (1990) Soil map of the world, revised legend. FAO, Rome, Italy

Friesen DK, Rao IM, Thomas RJ, Oberson A, Sanz JI (1997) Phosphorus acquisition and cycling in crop and pasture systems in low fertility soils. Plant Soil 196:289–294

Gilmore RM (1963) Fauna and ethnozoology of South America. In: Steward JH (ed) Handbook of South American Indians, vol 6. Physical anthropology, linguistics and cultural geography of South American Indians. Cooper Square, New York, pp 345–463

Glaser B (1999) Eigenschaften und Stabilität des Humuskörpers der "Indianerschwarzerden" Amazoniens. Bayreuther Bodenkundl Ber 68,196 pp

Hedley MJ, Stewart JWB, Chauhan BS (1982) Changes in inorganic and organic soil phosphorus fractions induced by cultivation practices and by laboratory incubations. Soil Sci Soc Am J 46:970–976

Kern DC, Kämpf N (1989) Antigos Assentamentos indigenas na formação de solos com terra preta arqueologica na região de Oriximina, Pará. Rev Bras Cienc Solo 13:219–225

Lehmann J, Cravo MS, Macedo JLV, Moreira A, Schroth G (2001a) Phosphorus management for perennial crops in central Amazonian upland soils. Plant Soil 237:309–319

Lehmann J, Günther D, Mota MS, Almeida M, Zech W, Kaiser K (2001b) Inorganic and organic soil phosphorous and sulfur pools in an Amazonian multistrata agroforestry system. Agrofor Syst 53:113–124

Lillios KT (1992) Phosphate fractionation of soils at Agroal, Portugal. Am Antiq 57:495–506

Mehlich A (1984) Mehlich 3 soil test extractant: a modification of the Mehlich 2 extractant. Commun Soil Sci Plant Anals 15:1409–1416

Mbahinzireki GB, Dabrowski K, Lee K-J, El-Saidy D, Wisner ER (2001) Growth, feed utilization and body composition of tilapia (*Oreochromis* sp.) fed with cottonseed meal-based diets in a circulating system. Aquacult Nutr 7:189–200

McCann JM, Woods WI, Meyer DW (2001) Organic matter and Anthrosols in Amazonia: interpreting the Amerindian legacy. In: Rees RM, Ball BC, Campbell CD, Watson CA (eds) Sustainable management of soil organic matter. CAB International, Wallingford, pp 180–189

Murphy J, Riley JP (1962) A modified single solution method for the determination of phosphate in natural waters. Anal Chim Acta 27:31–36

Naylor SJ, Moccia DR, Durant GM (1999) The chemical composition of settleable solid fish waste (manure) from commercial rainbow trout farms in Ontario, Canada. N Am J Aquacult 61:21–26

Petersen JB, Neves EG, Heckenberger MJ (2001) Gift from the past: *terra preta* and prehistoric Amerindian occupation in Amazonia. In: McEwan C, Barreto C, Neves EG (eds) Unknown Amazonia: culture in nature in ancient Brazil. British Museum, London, pp 86–105

Rathbone CK, Babbitt JK, Dong FM, Hardy RW (2001) Performance of juvenile Coho salmon *Oncorhynchus kisutch* fed diets containing meals from fish wastes, deboned fish wastes, or skin-and-bone by-products as the protein ingredient. J World Aquacult Soc 32:21–29

Rønsholdt B (1995) Effect of size/age and feed composition on body composition and phosphorus content of rainbow trout, *Oncorhynchus mykiss*. Water Sci Technol 31:175–183

Shimosaka C (1999) Relationship between chemical composition and crystalline structure in fish bone during cooking. J Clin Biochem Nutr 26:173–182

Smith NJH (1980) Anthrosols and human carrying capacity in Amazonia. Ann Assoc Am Geogr 70:553–566

Soil and Plant Analysis Council (1999) Soil analysis: handbook of reference methods. CRC Press, Boca Raton

Solomon D, Lehmann J (2000) Loss of phosphorus from soil in semi-arid northern Tanzania as a result of cropping: evidence from sequential extraction and ^{31}P-NMR. Eur J Soil Sci 51:699–708

Sombroek WG (1966) Amazon soils – a reconnaissance of soils of the Brazilian Amazon region. Centre for Agricultural Publications and Documentation, Wageningen, The Netherlands, 292 pp

Tiessen H, Moir JO (1993) Characterization of available P by sequential extraction. In: Carter MR (ed) Soil sampling and methods of analysis. Canadian Society of Soil Science, Lewis Publishers, Boca Raton, pp 75–86

Vacher S, Jérémie S, Briand J (1998) Amérindiens du Sinnamary (Guyane). Archéologie en forêt équatoriale. Documents d'Archéologie Française 70. Éditions de la Maison des Sciences de l'Homme, Paris, 297 pp

Wells EW, Terry RE, Parnell JJ, Hardin PJ, Jackson MW, Houston SD (2000) Chemical analyses of ancient anthrosols in residential areas at Piedras Negras, Guatemala. J Archaeol Sci 27:449–462

Woods WI (1977) The quantitative analysis of soil phosphate. Am Antiq 42:248–252

Woods W, McCann JM, Meyer DW (2000) Amazonian dark earth analysis: state of knowledge and directions for future research. In: Schoolmaster FA, Clark J (eds) Papers and Proc Geography Conf, vol 23, Tampa, Florida, pp 114–121

The Timing of *Terra Preta* Formation in the Central Amazon: Archaeological Data from Three Sites

EDUARDO GOÉS NEVES[1], JAMES B. PETERSEN[2],
ROBERT N. BARTONE[3], and MICHAEL J. HECKENBERGER[4]

9.1
Introduction

We present here chronological data on the timing and rate of *terra preta* formation at three archaeological sites located near the confluence of the Negro and Solimões Rivers, in the central Amazon of Brazil. We have been studying pre-Columbian indigenous archaeology in this area since 1995 within the framework of the Central Amazon Project (CAP) (e.g., Heckenberger et al. 1999; Neves 2000; Petersen et al. 2001). This research has identified more than 40 archaeological sites thus far, of which four (Açutuba, Osvaldo, Lago Grande, and Hatahara) have been tested and mapped in some detail (Fig. 9.1).

Stratigraphic excavation using 10-cm levels has been standard at the four most extensively studied sites, largely within natural stratigraphy in larger excavation units (e.g., Heckenberger et al. 1999; Neves 2000). These sites include the Hatahara, Lago Grande, and Osvaldo sites, discussed here, as well as the previously reported Açutuba site (Heckenberger et al. 1999). This chapter concentrates on the first three sites within the CAP study area – Hatahara, Lago Grande, and Osvaldo – where a large number of charcoal samples have been recently radiocarbon dated. In all three cases, we attempted to date at least one sample from each 10-cm level in order to obtain detailed chronological information on the timing of *terra preta* formation.

Our initial working hypothesis followed Nigel Smith's (1980) suggestion that a slow, long, and continuous process pertained to *terra preta* formation. Following Smith, we tentatively accepted that 1 cm of *terra preta* in the stratigraphy would be equivalent to about 10 years of intensive occupation. The results of radiocarbon dating for these three sites (and Açutuba) presented a different picture, however, indicating that *terra preta* formation in the CAP study area was at least sometimes a faster (and more complex) process than

[1] Museu de Arqueologia e Etnologia, Universidade of São Paulo, 1466 Ave. Prof. Almeida Prado, São Paulo 05508-900, Brazil
[2] Department of Anthropology, University of Vermont, Burlington, Vermont 05405, USA
[3] Archaeology Research Center, University of Maine at Farmington, Farmington, Maine 04938, USA
[4] Department of Anthropology, University of Florida, Gainesville, Florida 32611, USA

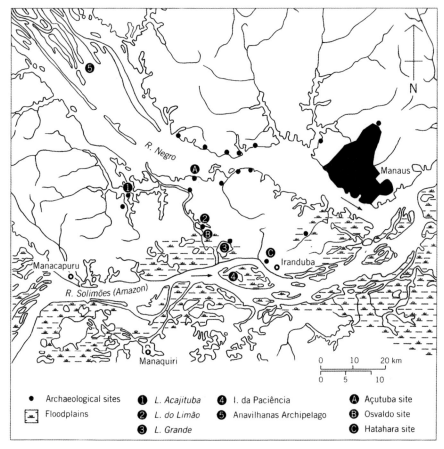

Fig. 9.1. The Central Amazon Project (CAP) study area

that predicted by Smith's hypothesis. We provide here a necessarily brief discussion of the procedures and results of fieldwork at the Hatahara, Lago Grande, and Osvaldo sites in order to substantiate this claim.

9.2
The Osvaldo Site

Osvaldo is a single occupation site with ceramics belonging to the Manacapuru phase from the so-called Barrancoid, or "Incised Rim," tradition (Hilbert 1968; Lathrap 1970; Heckenberger et al. 1999). Osvaldo is located on high ground along the southern shore of Lago do Limão and across from a contemporary village of the same name. This setting was chosen for study due to its location intermediate between the Negro and Solimões Rivers (Fig. 9.1). Since 1995, the CAP has been studying archaeological sites situated close to the rather different floodplains of the Negro and the Solimões. In 1999, we decided to sample sites located farther away from both of these large rivers. Lago do Limão is a tributary of the Ariaú River, a "furo," or variable flow channel, connecting the Negro and Solimões. Knowledge of this riverine passage is certainly pre-Columbian. Given its linear shape, it is likely that Lago do Limão is an ancient *igarapé* that was flooded as water levels in Amazonia rose during the Pleistocene–Holocene transition (Ab'Saber 1996).

There is information about other archaeological sites around Lago do Limão, but so far only Osvaldo has been studied. Vegetation across the site area is variable, composed of an old orange and lime orchard, small vegetable gardens, and thorny secondary growth, all growing on a dark, clay-rich *terra preta*. Osvaldo was delimited through the opening of a 1.0-km-long sampling transect over the landform. The content and depth of the archaeological deposits and *terra preta* were determined using a series of auger tests opened along the sampling transect at 25-m intervals. This sampling was complemented using a series of standard 0.5×0.5-m test pits, excavated in artificial 10-cm levels and spaced 100 m apart. In this way, for every 100 m along the sampling transect, one 0.5×0.5-m test pit and three auger tests were excavated. Three other transects were later opened, following the same procedures. In the end, 61 auger tests and 16 test pits were initially opened on and near Osvaldo. The site and broader area were also mapped using a total station laser transit to produce a detailed topographic map.

The auger tests and test pits across Osvaldo and beyond indicated that the depth of the *terra preta* and cultural deposits varied considerably within the site. Such variation has been interpreted as the archaeological record of a circular ring or horseshoe-shaped village, much like ethnographic and other known archaeological examples (e.g., Petersen 1996; Heckenberger et al. 1999). Based on these initial distribution and stratigraphic data, a single 1.0×1.0-m excavation unit was excavated in 10-cm levels within natural stratigraphy to a depth of about 100 cm. This excavation unit was situated in the area of a dense concentration of ceramic sherds and relatively deep *terra preta* on one side of the circular midden deposits.

Like other local sites, the *terra preta* was generally "black" in color (10YR2/1 in the Munsell soil color system) at Osvaldo and its base coincided almost exactly with the basal level of the dense ceramic deposits. The color of the sterile layer beneath the *terra preta* at the base of the excavation was "dark yellowish brown" (Munsell 10YR5/8), which was very different from the color of the overlying *terra preta* layers, again matching other local sites. The strong association of ceramics and *terra preta*, as well as the radical change in color, documents that the onset of *terra preta* formation was directly associated with the beginning of the Manacapuru occupation at Osvaldo.

After the fieldwork, three charcoal samples from the Osvaldo profile (samples 497, 498, and 498) were initially selected for radiocarbon dating. Twelve other samples were eventually also radiocarbon dated, all from beneath the depth of modern agricultural disturbance in the uppermost *terra preta*. Of these 15 radiocarbon samples from Osvaldo, 9 were recovered in situ, while 3 (samples 170, 471, and 474) were recovered during screening and thus may be less reliable. All these dated samples were derived from the single larger excavation unit at the Osvaldo site.

The 15 uncalibrated dates from Osvaldo, with one-sigma standard deviations, are presented in Table 9.1. They seemingly indicate that both *terra preta* formation and the Manacapuru occupation happened relatively quickly. Table 9.1 shows that the dates of the samples from essentially the bottom of the *terra preta* (samples 456 and 457) fall within the two-sigma standard deviation of the dates of the samples from the top of the *terra preta*, just below the cultivation zone (samples 498, 499, and 248).[5] Of course, this does not necessarily fully date the span of time represented by the *terra preta* at Osvaldo since the upper 30 cm or so of disturbed and partially disturbed deposits were not dated.

Generally speaking, if these interpretations are correct, we can assume that the bulk of the *terra preta*, perhaps more than 70 cm of the archaeological

Table 9.1. Osvaldo site, excavation unit S710 E1966, uncalibrated radiocarbon dates (1σ). All samples were charcoal fragments. Samples 497, 498, and 499 were recovered from the stratigraphic profile. Samples 170, 471, and 474 were recovered during screening. All other samples were recovered in situ during excavation. The date of sample 234 can be rejected because of its large standard deviation. Sample 167 presents a small inversion, but it falls within the standard deviation, while the date for sample 497 is rejected because of its aberrant position in the sequence

Sample no.	Depth	Date	Lab no.
	(cm)	(years B.P.)	
497	41	1550±40	Beta 143608
498	41	1290±30	Beta 143609
499	35	1370±40	Beta 143610
234	34	1730±90	Beta 143611
248	36	1330±40	Beta 143612
362	45	1290±40	Beta 143613
368	42	1360±50	Beta 143614
167	54	1440±70	Beta 143615
170	50–60	1350±30	Beta 143616
435	61	1340±40	Beta 143617
505	66	1350±40	Beta 143618
456	73	1320±60	Beta 143619
457	76	1310±40	Beta 143620
471	80–90	1980±80	Beta 143621
474	90–100	2120±40	Beta 143622

[5] The date of sample 234 can be rejected because of its large standard deviation. Sample 167 presents a small inversion, but it falls within the standard deviation, while the date for sample 497 is rejected because of its aberrant position in the sequence.

deposits, was formed within a century or so (±) in the seventh century A.D. (A.D. 601–700). However, it is also likely that some degree of disturbance and scant traces of possibly older and related occupation are represented at Osvaldo.

9.3
The Hatahara Site

The Hatahara site is much larger than Osvaldo, covering about 16 ha in terms of the *terra preta* and located on a bluff that overlooks a narrow floodplain strip along the northern bluff of the Solimões River near the town of Iranduba (Fig. 9.1). A combination of pasture, as well as papaya, manioc, and mango gardens covers the site surface and surrounding area. Testing and mapping procedures were largely similar to those described above for Osvaldo. Sampling transects were also used to sample Hatahara in 1999, but in this case only auger tests were initially used to sample the spatial variability and depth of the *terra preta* and other cultural deposits. Other recent excavations were conducted at Hatahara in 2001 and 2002, but these are not reported here. Hatahara preserves evidence of three more or less distinct occupations: a 70-cm-thick basal Manacapuru phase (Barrancoid tradition) stratum with transitional *terra preta* (probably deposited from overlying strata) covered by thick *terra preta* deposits assigned to sequential Paredão and Guarita occupations (Amazon Polychrome tradition), respectively (Hilbert 1968; Lathrap 1970).

As at other sites in the CAP study area, several mounds are represented at Hatahara and these are variably visible on the surface where the vegetation has been cleared. The data presented here were collected in 1999 through the excavation sampling of one of these mounds using several contiguous 1.0×1.0-m excavation units, a funerary structure containing three composite secondary burials with at least ten individuals.

Radiocarbon dates for the initial 2.0-m-deep excavation at Hatahara are presented in Tables 9.2 and 9.3. The dates listed in Table 9.2 display apparent aberrations that can be only understood if one recognizes disturbances related to emplacement of the three burials and mound construction over them. Therefore, inversions on the dates of samples 549, 589, 1840, and 1848 likely resulted from relocation of cultural deposits when the burials disturbed the older, deeper cultural deposits. Samples 222, 394, 1882, and 1860 are also problematic, either because their dates lie outside of the sequence or because of their relatively large standard deviations. Table 9.3 presents a "clean," selected interpretation of the sequence from the 1999 units at Hatahara, eliminating the problematic suspect dates.

The corrected, selected Hatahara sequence in Table 9.3 shows that *terra preta* formation was closely associated with the Guarita and Paredão occupations. Preliminary interpretation indicated that *terra preta* probably first formed during the initial Paredão occupation, being later recycled as raw

Table 9.2. Hatahara site, excavation unit N1152 W1360, uncalibrated radiocarbon dates (1σ). All samples were charcoal fragments. Samples 361, 394, 594, 1848, 1855, 1860, and 1869 were recovered during screening. Samples 1879, 1880, 1882, and 1881 were recovered from the profile. All other samples were recovered in situ during excavation

Sample no.	Depth	Date	Lab no.
	(cm)	(years B.P.)	
222	29	350±40	Beta 143582
361	30–40	1010±80	Beta 143583
394	40–50	1250±70	Beta 143584
1879	58	980±40	Beta 143585
1880	60	960±30	Beta 143586
1882	65	570±40	Beta 143587
1881	80	1000±40	Beta 143588
505	84	1000±40	Beta 143589
549	100–110	1250±80	Beta 143591
589	121	910±40	Beta 143592
1840	130	1070±50	Beta 143593
1848	140–150	890±120	Beta 143594
1892	155	960±40	Beta 143595
1855	160–170	1070±70	Beta 143596
1860	170–180	2310±120	Beta 143597
1869	180–190	1080±40	Beta 143598
1873	192	1300±40	Beta 143599

Table 9.3. Hatahara site, selected/revised radiocarbon sequence, uncalibrated (1σ)

Sample no.	Depth	Date	Lab no.
	(cm)	(years B.P.)	
361	30–40	1010±80	Beta 143583
1879	58	980±40	Beta 143585
1880	60	960±30	Beta 143586
1881	80	1000±40	Beta 143588
505	84	1000±40	Beta 143589
1892	155	960±40	Beta 143595
1855	160–170	1070±70	Beta 143596
1869	180–190	1080±40	Beta 143598
1873	192	1300±40	Beta 143599

material for mound building during either the later Paredão occupation or the Guarita occupation. Regarding this chronology of *terra preta* formation, the selected sequence indicates the same general pattern as at Osvaldo: one of a very rapid building (note the overlapping standard deviations for samples 505, 361, 1879, 1880, 1881, and 505). This is not completely surprising, however, given the partial attribution of cultural deposits in this particular area to intentional mound building.

This sequence also shows a somewhat puzzling circumstance: *terra preta* was largely formed during the relatively brief Guarita and Paredão occupations, with little formed during the apparently longer occupation of the underlying Manacapuru phase (samples 1873, 1869, and 1855). However, this observation may be the result of particular local conditions and may or may not be reflective of the site as a whole. In any case, the depth and density of this particular *terra preta* formation is not directly related to the full length of occupation at Hatahara, especially since we sampled a mound (also see Heckenberger et al. 1999). These data point to the fact that single excavations in large *terra preta* sites may not be reflective of the entire site and we must try to recognize spatial and temporal variations in *terra preta* and other site formation processes whenever possible.

9.4
The Lago Grande Site

This site is located on a bluff on the north shore of a floodplain lake, most likely an abandoned meander, connected to the Solimões River, and relatively close to its northern border (Fig. 9.1). The linear distance of the site to the river proper is around 5 km, attesting to Denevan's (1996) "bluff model" for settlement location along the Amazon River and its major tributaries. Other sites, not yet studied, are also located on the northern shore of Lago Grande. At the Lago Grande site, the *terra preta* is covered by small areas of cultivation, but mostly by dense secondary forest growth. A defensive ditch is present on the landward side of the site landform. Analogous to Hatahara, several subtle mounds are visible at Lago Grande, but no burials have been identified yet. Data presented here were gathered in the test excavation of one of these mounds.

Somewhat different from the other two cases, the darkest *terra preta* layer at Lago Grande was not found at the surface, but rather it occurs in stratum III at a depth of around 90 cm below the modern surface. Stratum III is also correlated with the highest density of pottery sherds in the excavation profile, in this case with ceramics of the Paredão phase like those at Hatahara (Hilbert 1968).

Dates obtained for stratum III at Lago Grande (samples 330, 329, 322, and 325) indicate that the occupation of the deposit lasted several hundred years (±), starting at around A.D. 700/800 and lasting until approximately A.D. 1000 (Table 9.4). The data show a relatively fast initial building process, in part 100 years or less for 40 cm of *terra preta*, but again this may be due to some part of the mound construction. It remains unclear whether or not these data reflect the full span of occupation at Lago Grande and the true nature of the cultural deposits all across the site, since again the uppermost deposits were not dated to avoid possible historic disturbance due to agriculture.

Table 9.4. Lago Grande site, excavation unit 1, uncalibrated radiocarbon dates (1σ). All samples were charcoal fragments recovered from the profile

Sample no.	Depth	Date	Lab no.
	(cm)	(years B.P.)	
319	36	1050±40	Beta 143600
324	75	950±40	Beta 143601
326	83	950±30	Beta 143602
321	89	960±30	Beta 143607
325	118	1130±40	Beta 143604
322	123	1150±40	Beta 143603
329	142	1100±30	Beta 143605
330	158	1260±40	Beta 143606

9.5
Conclusions

If the interpretations presented here are reliable, they indicate that *terra preta* formation in central Amazonia was a more variable and sometimes faster process than previously thought. At the Hatahara and Lago Grande sites, there is no correlation between length of occupation and *terra preta* formation, at least as related to mound construction, which reflects complicated, differential episodic and likely idiosyncratic formation of cultural deposits, with some long-term continuous deposits represented as well. In contrast, at Osvaldo there were no mounds, and yet a combination of continuous and episodic formation deposition of cultural deposits seems likely there too. If this is the case, what social processes can be postulated to explain these phenomena?

We suggest that particular, sometimes very localized, intensive activities and, more broadly, *population density* rather than *time* alone, should be key factors explaining *terra preta* formation. In this sense, the development of *terra preta* at sites with multiple occupations can be seen as an archaeological correlate of population growth. Such an hypothesis can be tested with further fieldwork in the CAP study area and other settings where *terra preta* sites occur. These CAP examples also clearly document that individual sites will not be necessarily well understood through limited, localized, small-scale excavations. In fact, the excavations reported here and others at the Açutuba site well demonstrate the potential variability in spatial and stratigraphic contexts across individual *terra preta* (and other) archaeological sites, sometimes with extreme variation over very limited distances. We cannot study these sites easily, nor can we easily generalize from single excavations or single soil profiles of them. This is a very important point. These sites clearly need and deserve broad-scale sampling and archaeologists and soil scientists alike should be cognizant of this fact.

We feel that our research can be also used to criticize theories of broad-brush environmental determinism in Amazonia. If large population aggregates produced the settlements that we study as *terra preta* sites in the central Amazon and elsewhere, limiting environmental factors to population growth may not be very relevant, or relevant at all, in this region. In this sense, other factors should be sought to explain settlement formation, expansion, and abandonment regionally. Defensive ditches at several sites within the CAP area and elsewhere (e.g., Heckenberger 1996; Heckenberger et al. 1999; Neves 1998, 2000, 2001) support the idea that warfare was widespread by at least the last millennium of pre-Columbian occupation in Amazonia (A.D. 500–1500), long before the beginning of the European colonization. Further research is needed to study the remnants of pre-Columbian settlements in central Amazonia for this and various other reasons.

Acknowledgements. Fieldwork and radiocarbon dating of the Hatahara, Lago Grande, and Osvaldo sites were funded by grant 99/02150-0 from the Fundação de Amparo Pesquisa do Estado de São Paulo (FAPESP), and initial CAP funding was provided by various other individuals, institutions, and organizations, including the Wenner-Gren Foundation, the William T. Hillman Foundation, and the University of Maine at Farmington Archaeology Research Center, among others (see Heckenberger et al. 1999; Petersen et al. 2001). In Manaus, we warmly thank Profs. Francisco Jorge dos Santos and Luis Balkar Sá Peixoto Pinheiro, Directors of Museu Amazônico, Universidade Federal do Amazonas, and Ms. Ana Lúcia Abrahim, head of the 1ª Superintendência of IPHAN (Instituto do Patrimônio Histórico e Artístico Nacional), for their continued support. We also thank Mr. Osvaldo Gomes da Silva, Mr. José Ricardo, Mr. José Mitonho, Mr. Adilson Santos, Mr. Yodi Ideta, and Mr. Kuni-taka Ideta for kindly allowing us to work on their properties. We thank all of our students and other crew members for their help in both field and laboratory. Drawings and maps were done by Marcos Castro. Finally, we thank Bill Woods for his help with preparation of this chapter.

References

Abŝaber A (1996) Paleoclima e paleoecologia da Amazônia Brasileira. In: A Amazônia: Do discurso à praxis. Editora da Universidade de São Paulo, São Paulo, pp 49–66

Denevan WM (1996) A bluff model of riverine settlement in prehistoric Amazonia. Ann Assoc Am Geogr 86(4):654–681

Heckenberger M (1996) War and peace in the shadow of empire: sociopolitical change in the Upper Xingu of southeastern Amazonia, A.D. 1400–2000. PhD Diss, Department of Anthropology, University of Pittsburgh

Heckenberger MJ, Petersen JB, Neves EG (1999) Village size and permanence in Amazonia: two archaeological examples from Brazil. Latin Am Antiq 10(4):533–576

Hilbert PP (1968) Archäologische untersuchungen am Mittlern Amazonas: Beiträge zur vorgeschichte des südamerikanischen tieflandes. Dietrich Reimer, Berlin

Lathrap DW (1970) The Upper Amazon. Thames and Hudson, London

Neves EG (1998) Paths in dark waters: archaeology as indigenous history in the Upper Rio Negro Basin, northwest Amazon. PhD Diss, Department of Anthropology, Indiana University

Neves EG (2000) Levantamento arqueológico da área de confluência dos rios Negro e Solimões, Estado do Amazonas, relatório de atividades Junho 1999–Agosto 2000. Report submitted to the Fundação de Amparo à Pesquisa do Estado de São Paulo (FAPESP)

Neves EG (2001) Indigenous historical trajectories in the Upper Rio Negro Basin. In: McEwan C, Barreto C, Neves EG (eds) Unknown Amazon: culture in nature in ancient Brazil. British Museum Press, London, pp 266–286

Petersen JB (1996) The archaeology of Trants (Monserrat), Pt. 3. Chronological and settlement data. Ann Carnegie Mus 65:323–361

Petersen JB, Neves EG , Heckenberger MJ (2001) Gift from the past: *terra preta* and prehistoric Amerindian occupation in Amazonia. In: McEwan C, Barreto C, Neves EG (eds) Unknown Amazon: culture in nature in ancient Brazil. British Museum Press, London, pp 86–105

Smith NJH (1980) Anthrosols and human carrying capacity in Amazonia. Ann Assoc Am Geogr 70(4):553–566

10 Semi-Intensive Pre-European Cultivation and the Origins of Anthropogenic Dark Earths in Amazonia

WILLIAM M. DENEVAN[1]

10.1
Introduction

Anthropogenic dark earths are widespread in the uplands (*terra firme*) of Amazonia, in patches covering a hectare or less up to several hundred hectares. The blacker form (*terra preta*) seems to have developed from pre-European village middens consisting of ash and charcoal from kitchen fires, cultural debris, feces, human and animal bones, and house/garden waste (Woods and McCann 1999). The lighter, dark brown form (*terra mulata*), which is much more extensive, is believed by some soil scientists (Sombroek 1966:175; Glaser et al. 2001a), archaeologists (Herrera et al. 1992; Petersen et al. 2001), botanists (Prance and Schubart 1978), and geographers (Denevan 1998; Woods and McCann 1999) to be the product of intensive cultivation practices (Fig. 10.1). Others, however, such as Smith (1980) and Eden et al. (1984), rejected an agricultural origin because of the depth of dark earth soils. Smith argued for midden origins, and he saw soil color and depth as being functions of length of village-site duration. In 1980 he was apparently unaware of the extent of *terra mulata* that contains little or no midden material.

Dark earth soils in Amazonia are considerably higher in fertility (pH, P, Mg, Ca, organic matter) than are surrounding soils. Color and chemical characteristics have been attributed by Glaser et al. (2001a, b, 2002) to a high content of black carbon particles that are residues of incomplete combustion from frequent burning by pre-European people.[2] Incomplete combustion, producing charcoal instead of ash, can be accomplished by burning when the cut woody vegetation is still moist rather than well into the dry season, a 'cool' burn in contrast to a 'hot' burn.[3] The necessary burning frequency potentially could have been provided in zones of semi-intensive cultivation[4] surrounding permanent villages. What were the cultivation practices that

---

[1] University of Wisconsin-Madison, P.O. Box 853, Gualala, California 95445, USA
[2] Natural fires in rain forest in the northern Amazon in the mid-Holocene only occurred at a site frequency of 389–1,540 years (Kauffman and Uhl 1990).
[3] Average carbon recovery from charred woody biomass ('slash-and-char') is 50% compared to only 3% with common slash-and-burn practices (Glaser et al. 2002).
[4] In semi-intensive (or semi-permanent) cultivation, both the cultivation and fallow periods are brief (from one to several years)

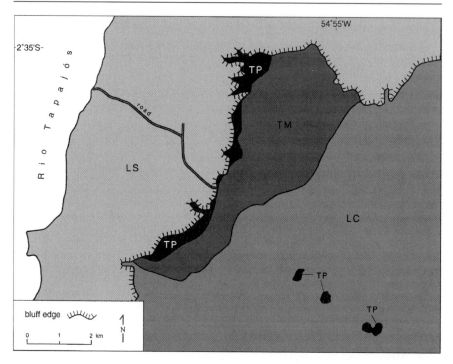

Fig. 10.1. Map showing bluff-edge *terra preta* (ca. 200 ha) along the Belterra Plateau adjacent to the Rio Tapajós, Brazilian Amazonia, with *terra mulata* (ca. 1,000 ha) inland. Note the small patches of *terra preta* beyond the bluff zones. *TP tierra preta*; *TM terra mulata*; LS latosolic sand (sloping, eroded, plateau front); LC latosolic clay (interior plateau). (Adapted from Sombroek 1966:175; Denevan 2001:106)

could have produced frequent burning and carbon-rich and organic-matter-rich dark earth soils? The focus here is on the *terra mulata* soils.

10.2
Potential Origin of *Terra Mulata*

We know that *terra mulata* generally is not formed by long-fallow shifting cultivation today (Pabst 1993). A lengthy period of time apparently is required, not a few years but rather decades or even centuries. If, indeed, this soil is related to cultivation activity, then some form of semi-intensive cropping involving frequent burning must have been involved.

10.2.1
Frequent Burning of Cleared Short Fallow

A short cropping/fallow cycle of cultivation probably was critical. In the lower Rio Negro region of Brazil today, a single year of cropping of dark earth soil on *terra firme* is followed by 4–5 years of fallowing (German 2003a, b), what can be considered a semi-intensive system with frequent reclearing and burning. Because of the high fertility of dark earths in Amazonia, weed invasion is very aggressive immediately after clearing, burning, and planting.[5] However, this is countered by only a few years of fallowing. Weeding requirements are greater after short fallows than after long fallows, but are still less than after only 1 or 2 years of cultivation, and they are progressively greater as cultivation continues. On fertile dark earth soils, fallowing is probably more important for weed management than for restoration of fertility, but this is also often true on other soils (Chagnon 1992).

I have argued (Denevan 1992) that long-fallow shifting cultivation, so common today by both Indians and non-Indians, was infrequent in pre-European times because of the inefficiency of stone axes for clearing forest. The introduction of iron axes and, later, steel axes by Europeans induced a major, but negative, agricultural revolution by making possible labor-efficient, but land extensive, long-fallow shifting cultivation with frequent clearing for new fields. Previously, once a clearing for crops was available, it could have been farmed frequently in a short cycle similar to that reported by German (2003a). Fertility can be maintained by mulching, composting, intercropping, in-field burning, and other traditional techniques still practiced by some Amazonian Indians (Hecht 2003). Natural clearings are created by tree falls, wild fires, and tree blow-downs during violent storms (Denevan 2001:120, 125). Artificial clearings could have been made with stone and wood tools where vegetation was easy to clear. Once established, a clearing could have been gradually enlarged.

Short-lived cultivation plots likely were located within a mosaic of young, managed second growth which included fruit trees, dispersed or in small orchards. This growth probably was similar to the 5- and 6-year-old managed orchard fallows that we mapped and described for the Bora in northern Peruvian Amazonia (Denevan and Treacy 1988). The importance of fruit and nut trees on dark earths in pre-European food systems is known from archaeology at Araracuara on the Río Caquetá in Colombian Amazonia (Herrera et al. 1992) and historically from Spanish chroniclers in the mid-sixteenth century (Denevan 1996). Thus, the burning of young fallows for new clearings could have occurred with a frequency of only a few years.

10.2.2
In-field Burning

"In-field burning" refers to burning within a crop field or agroforestry zone, subsequent to field establishment. Particularly common today are secondary burns in which charred trunks and branches, not initially completely burned,

---

[5] In the Rio Arapiuns region west of Santarém, on dark earths there may be three weedings per crop compared to one or none on adjacent oxisols (McCann 2004).

are piled and reburned both to clear space and to create ash piles in which crops may be concentrated. Other burn piles consist of weeds, crop residues, and transported forest litter. In addition to concentrating nutrients and reducing soil acidity, burning reduces insects and disease pathogens.

The farming practices of the Kayapó of central Brazil are an example of carefully controlled in-field burning, as is described by Hecht and Posey (1989:180; also see Hecht 2003). For the first 3 years of cultivation, the Kayapó manage their small burns for frequency, location, and extent, and for "volume of biomass, seasonal timing, diurnal timing, and [thus] the temperature of the burn," and these are related to specific crops and crop clustering. Burning patterns are important for the structure or architecture of Kayapó fields. The species burned can also "affect the fertility characteristics of the burn." The Kayapó have many descriptive terms for types of ash, as well as songs and rituals about burning, and their shamans specialize in burning techniques.

Some Amazonian people spend only a total of one or a few hours on burning and reburning a clearing (Barí, Kuikuru, Mirití, Shipibo, Amahuaca), whereas the Machiguenga, who are very careful burners, average 80 hours (Beckerman 1987). Other people besides the Kayapó, such as the Machiguenga, believe that ash is a good fertilizer for crops (Johnson 1983). Generally, however, a layer of ash is only a temporary soil enrichment and may not be so recognized at all by Indian farmers (Descola 1994).

Another possibility for regular in-field burning is the cutting of tree branches, piling, drying, and then burning them to create ash-charcoal piles. These could be branches from managed trees, secondary growth, or primary forest, with the trees generally recovering from such branch removal. There is little evidence for this either historically or at present in Amazonia, but it is a possibility and, it could have been done with stone axes and *macanas* (hard chonta-palm machetes). As a model for such branch lopping, we can look at the *citemene* system in central Africa (Allan 1965; Stromgaard 1985, 1988; Oyama 1996; Quinby 2001).

The Bemba and other people in Zambia lop off trunks and branches of woodland trees and stack and burn them in either small circles 20–30 ft in diameter or in large circles of as much as an acre.[6] In these otherwise poor soil regions, crops are planted for up to 5 years, followed by 20 years or so of tree recovery before they are lopped again. The regional *citemene* landscapes are covered by these circles in various stages. Phosphate, potassium, calcium, and pH levels are raised by the ash residues. Apparently the "hot" fires in this process, producing mainly ash, do not result in a permanent dark, fertile, anthropogenic soil. "Cool" burning leaving charcoal potentially could do so. Population densities of 4–16 per square mile (Quinby 2001) have been supported by this system for long periods, which is far larger than known Indian densities of less than 1/square mile in upland Amazonia.

---

[6] For the large circles, the ratio of size of the cropped area to the cut area is 1:6.6 (Oyama 1996).

The Achuar in Ecuador (Descola 1994), shortly after the initial burn, gather incompletely consumed branches, and in the center of the clearing they make a large stack that is then burned. Apparently, however, branches are not brought in from outside the clearing. Smaller piles of slash are arranged around stumps and burned, with the resulting ash concentrations being favored for planting yams that do well in potassium-rich soils.

10.2.3
Organic Amendments

Woods and McCann (1999) believe that a combination of frequent burning and organic inputs from mulching and composting resulted in heightened nutrient-retaining capacity, enhanced fertility, and self-perpetuating soil biotic activity, leading to the formation and persistence of *terra mulata*. Also, unburned slash can be left to decompose relatively rapidly from humidity and high temperatures, slowly releasing nutrients.

Kayapó mulching and composting are very important for soil management and include crop residues, chopped weeds, banana leaves, and also palm fronds that in this case may be brought from the adjacent forest (Hecht and Posey 1989; Hecht 2003). In addition, there may be direct applications of organic material mixed into the soil around tree crops (ash, ant and termite nests, bones, shredded leaves). With these practices and in-field burning, the Kayapó are able to crop for about 5 years, followed by only 10–11 years of fallow. Fertility indicators (pH, N, P, K, Ca, Mg) not only actually increased (over forest levels) for the first year following initial burning, but also were generally sustained in the fifth year of cropping and even in the tenth year (Hecht 1989). Abandonment after the fifth year was due to intensified weed invasion, not fertility decline. There are similar descriptions for other Indians in Amazonia. The Waika (Yanomami group) in Venezuela generally cultivate polycultural swiddens for 5 or 6 years (Harris 1971). At Araracuara, there is archaeological evidence that fertility was improved by the addition of "[o]rganic material, such as domestic waste, dead leaves, wood, and weeds" to the soil (Mora et al. 1991:79).

Where dry periods are too short for burning, all cut vegetation may be left on the ground to decompose without burning. Such "slash/mulch" systems have been described by Thurston (1997). There are several examples today in western Amazonia (Achuar, Canelos Quichua, Napo Quichua), and the technique may have been more common in prehistory. The Kayabí Indians in Brazil both burn slash and allow unburned material to decompose (Rodrigues 1993).

In the lower Rio Negro region, *caboclo* farmers today counter fertility loss by adding nutrients to both dark earth and common soils by burning weed and leaf residues, by leaving residues in place to decompose, and by making large compost piles (German 2003).

10.2.4
Permanent Settlement and Dark Earths

The evidence for semi-intensive pre-European cultivation is in part inferential, and the dark earths themselves are such evidence. Furthermore, we now know that villages were not necessarily small, a few hundred people or less, but rather some numbered in the thousands, and these villages were not necessarily shifted but were in some places permanent. There is ethnohistorical evidence for both such size and permanence (Denevan 2003). Furthermore, there is archaeological evidence for such villages: at Araracuara (Mora et al. 1991; Herrera et al. 1992); in the lower Rio Negro region (Petersen et al. 2001); and in the upper Xingu region (Heckenberger 1998; Heckenberger et al., 1999). Heckenberger mapped two Xingu sites that were 40 and 50 ha in size, most of this containing dark earth soils. In comparison, nearby present-day Kuikuru villages are 5 ha or less in area. The former were occupied continuously from at least A.D. 1000–1500 and contained permanent earthworks including multiple circular moats, roads, and embankments, with estimated populations of 1,500 or more. Many more such large pre-European village sites undoubtedly will be found under upland forest.[7].

There is even greater evidence for large sites on *terra firme* river bluffs along the Amazon River and its major tributaries (Denevan 1996). Permanent villages had to have been sustained mainly by frequent cultivation, and many of those studied by archaeologists are associated with anthropogenic dark earth soils.

I should emphasize that archaeological village sites were not necessarily fully occupied at one point in time. Houses were likely shifted around within a site because of decaying structures, buildup of vermin and garbage, or a death. Even today, there are instances of Yanomamö villages staying in one area for 60–80 years with micro-movement within the area (Chagnon 1992). Some Kuikuru have lived in the same area for 90 years, moving their villages four times but only a few hundred yards apart (Carneiro 1961). Thus locational advantages were retained, including access to water, to improved soils, and to managed or manipulated plants. Meggers (2001), however, argues that entire pre-European sites were not occupied continuously but rather were repeatedly abandoned and reoccupied by small groups, thereby accounting for large occupation areas and midden depths. DeBoer et al. (2001), Heckenberger et al. (2001), and other archaeologists disagree.

---

[7] See Pärssinen and Korpisaari (2003) on discoveries of extensive pre-European earthworks in recently deforested areas of Acre, Brazilian Amazonia.

10.3
Conclusions

The scenario I suggest here is that:

1. Given the inefficiency of stone axes for clearing mature forest, pre-European farmers often relied on short-cropping/short-fallow systems, a cycle of 1 or 2 years of cultivation and 4–5 years of fallow, or by 4 or 5 years of cropping and 10 years or so of fallow, or by other similar variations, with the fallows being managed agroforestry systems dominated by fruit trees and other useful species. The short fallows would have been insufficient to rejuvenate soil fertility which instead was sustained by ash and charcoal from frequent burning of fallow and in-field burning, and by composting and mulching. These were patches of various sizes that were fairly permanent, but in which there was rotation of cultivated fields with managed fallows and fruit orchards.

2. Subsequently, over considerable time, given concentrated, semi-continuous cultivation activity, dark-brown *terra mulata* soils were formed in the persistent, frequently burned cultivation–agroforestry zones. Once established, these superior dark earth soils were particularly attractive to farmers because nutrient-demanding crops such as maize could be grown successfully on them. Thus, these soils acted as further stimulus for both maintaining settlement and cultivation in these sites and for returning to them when/if there were periods of abandonment, continuing to the present. Continuity over hundreds of years is indicated for some interior upland sites and riverine bluff sites. The preceding portrayal, however, is an oversimplification and is not intended to be universal. Undoubtedly, there has been considerable variation in Amazonia in the forms of dark earths, the specific processes and histories responsible, forms of land use, and associated settlement patterns.

In conclusion, I argue, with others, that it is possible that many dark earth soils could have been the product of semi-intensive agricultural practices. This is conjecture, but I have tried to support it with examples of current Indian practices in Amazonia. Unfortunately, the process of dark earth soil formation will be difficult to replicate given the long time spans apparently involved. However, analyses of soil chemistry and micromorphology and plant microfossils should be instructive as to soil genesis and cultivation activity.

References

Allan W (1965) The African husbandman. Barnes and Noble, New York
Beckerman S (1987) Swidden in Amazonia and the Amazon rim. In: Turner II BL, Brush SB (eds) Comparative farming systems. Academic Press, New York, pp 55–94
Carneiro RL (1961) Slash-and-burn cultivation among the Kuikuru and its implications for cultural development in the Amazon Basin. In: Wilbert J (ed) The evolution of horticultural systems in native South America. Sociedad de Ciencias Naturales La Salle, Caracas, pp 47–67

Chagnon NA (1992) Yanomamö: the last days of Eden. Harcourt Brace, San Diego, pp 91–92

DeBoer WR, Kintigh K, Rostoker AG (2001) In quest of prehistoric Amazonia. Latin Am Antiquity 12:326–327

Denevan WM (1992) Stone vs metal axes: the ambiguity of shifting cultivation in prehistoric Amazonia. J Steward Anthropol Soc 20:153–165

Denevan WM (1996) A bluff model of riverine settlement in prehistoric Amazonia. Ann Assoc Am Geogr 86:654–681

Denevan WM (1998) Comments on prehistoric agriculture in Amazonia. Culture Agric 20:54–59

Denevan WM (2001) Cultivated landscapes of native Amazonia and the Andes. Oxford University Press, Oxford

Denevan WM (2003) The native population of Amazonia in 1492 reconsidered. Rev Indias (Madrid) 63:175–187

Denevan WM, Treacy JM (1988) Young managed fallows at Brillo Nuevo. Adv Econ Bot 5:8–46

Descola P (1994) In the society of nature: a native ecology in Amazonia. Cambridge University Press, Cambridge, pp 157–158

Eden MJ, Bray W, Herrera L, McEwan C (1984) *Terra preta* soils and their archaeological context in the Caquetá Basin of southeast Colombia. Am Antiquity 49:125–140

German LA (2003a) Historical contingencies in the coevolution of environment and livelihood: contributions to the debate on Amazonian black earths. Geoderma 111:307–331

German LA (2003b) Ethnoscientific understandings of Amazonian dark earths. In: Lehmann J, Kern DC, Glaser B, Woods WI (eds) Amazonian dark earths: origin, properties, management. Kluwer, Dordrecht, pp. 179–201

Glaser B, Haumaier L, Guggenberger G, Zech W (2001a) The '*terra preta*' phenomenon: a model for sustainable agriculture in the humid tropics. Naturwissenschaften 88:37–41

Glaser B, Guggenberger G, Zech W (2001b) Black carbon in sustainable soils of the Brazilian Amazon region. In: Swift RS, Spark KM (eds) Understanding and managing organic matter in soils, sediments, and waters. International Humic Substances Society, St. Paul, pp 359–364

Glaser B, Lehmann J, Zech W (2002) Ameliorating physical and chemical properties of highly weathered soils in the tropics with charcoal: a review. Biol Fertil Soil 35:219–230

Harris DR (1971) The ecology of swidden cultivation in the Upper Orinoco rain forest. Geogr Rev 61:475–495

Hecht SB (1989) Indigenous soil management in the Amazon Basin: some implications for development. In: Browder JO (ed) Fragile lands of Latin America. Westview Press, Boulder, pp 166–181

Hecht SB (2003) Indigenous soil management and the creation of *terra mutata* and *terra preta* in the Amazon Basin. In: Lehmann J, Kern DG, Glaser B, Woods WI (eds) Amazonian dark earths: origin, properties, management. Kluwer, Dordrecht, pp 355–372

Hecht SB, Posey DA (1989) Preliminary results on soil management techniques of the Kayapó Indians. Adv Econ Bot 7:174–188

Heckenberger MJ (1998) Manioc agriculture and sedentism in Amazonia: the upper Xingu example. Antiquity 72:633–648

Heckenberger MJ, Petersen JB, Neves EG (1999) Village size and permanence in Amazonia: two archaeological examples from Brazil. Latin Am Antiquity 10:353–376

Heckenberger MJ, Petersen JB, Neves EG (2001) Of lost civilizations and primitive tribes, Amazonia: reply to Meggers. Latin Am Antiquity 12:328–333

Herrera LF, Cavelier I, Rodriguez C, Mora S (1992) The technical transformation of an agricultural system in the Colombian Amazon. World Archaeol 24:98–113

Johnson A (1983) Machiguenga gardens. In: Hames RB,Vickers WT (eds) Adaptive responses of native Amazonians. Academic Press, New York, pp 29–63

Kauffman JB, Uhl C (1990) Interaction with anthropogenic activities, fire, and rain forests in the Amazon Basin. In: Goldammer JG (ed) Fire in the tropical biota: ecosystem processes and global challenges. Ecological studies 84. Springer, Berlin Heidelberg New York, pp 117–130

McCann JM (2004) Subsidy from culture: anthropogenic soils and vegetation in Tapajônia, Brazilian Amazonia. PhD Diss, University of Wisconsin-Madison

McCann JM, Woods WI, Meyer DW (2001) Organic matter and anthrosols in Amazonia: interpreting the Amerindian legacy. In: Rees RM, Ball BC, Campbell CD, Watson CA (eds) Sustainable management of soil organic matter. CAB International, New York, pp 180–189

Meggers BJ (2001) The continuing quest for El Dorado: round two. Latin Am Antiquity 12:304–325

Mora S, Herrera LF, Cavelier I, Rodriguez C (1991) Cultivars, anthropic soils and stability: a preliminary report of archaeological research in Araracuara, Colombian Amazonia. Latin American archaeology reports no. 2. University of Pittsburgh, Pittsburgh

Oyama S (1996) Regeneration process of the Miombo woodland at abandoned fields of *citemene* shifting cultivation in northern Zambia. Afr Study Monogr 17:101–116

Pabst EE (1993) *Terra preta*: Ein beitrag zur genesediskussion auf der basis von geländearbeiten bei Tupí-Völkern Amazoniens. Diss, Universität Kassel, Kassel

Pärssinen M, Korpisaari A (eds) (2003) Western Amazonia – Amazônia Ocidental. Renvall Institute, University of Helsinki

Petersen JB, Neves EG, Heckenberger MJ (2001) Gift from the past: *terra preta* and prehistoric Amerindian occupation in Amazonia. In: McEwan C, Barreto C, Neves EG (eds) Unknown Amazon: culture in nature in ancient Brazil. British Museum Press, London, pp 86–105

Prance GT, Schubart HOR (1978) Notes on the vegetation of Amazonia. 1: A preliminary note on the origin of the open white sand campinas of the lower Rio Negro. Brittonia 30:60–63

Quinby EG (2001) Merging GIS/RS and socio-environmental analytical methods in the study of land-use change: the case of *citemene* agriculture in the northern province of Zambia. Master's Thesis, University of Wisconsin-Madison, p 137

Rodrigues AJ (1993) Ecology of the Kayabí Indians of Xingú, Brazil: soil and agroforestry management. Doctoral Diss, University of Cambridge, Cambridge, pp 154–155

Smith NJH (1980) Anthrosols and human carrying capacity in Amazonia. Ann Assoc Am Geogr 70:553–566

Sombroek WG (1966) Amazon soils: a reconnaissance of the Brazilian Amazon region. Centre for Agricultural Publications and Documentation, Wageningen

Stromgaard P (1985) The infield–outfield system of shifting cultivation among the Bemba of south-central Africa. Tools Tillage 5(2):67–84

Stromgaard P (1988) Soil and vegetation changes under shifting cultivation in the Miombo of East Africa. Geogr Ann 70B:303–374

Thurston HD (1997) Slash/mulch systems: sustainable methods for tropical agriculture. Westview Press, Boulder

Woods WI, McCann JM (1999) The anthropogenic origin and persistence of Amazonian dark earths. Yearb Conf Latin Am Geogr 25:7–14

11 Identifying the Pre-Columbian Anthropogenic Input on Present Soil Properties of Amazonian Dark Earths (Terra Preta)

Bruno Glaser[1], Georg Guggenberger[2], and Wolfgang Zech[1]

11.1 Introduction

Soil degradation is one of the most serious problems of land use in the tropics, in particular the Neotropics, because the majority of the soils are very old, deeply weathered, and rather infertile (Zech 1997). Under tropical conditions, hydrolysis of primary minerals, for instance, is enhanced by a factor of 100 in comparison to temperate regions (Whitmore 1993). In combination with the high precipitation, the released nutrients are highly susceptible to leaching. Clearing of forests or woodlands and their conversion into farmland reduces the soil organic matter (SOM) contents due to higher mineralization rates and reduced litter input and increases soil erosion (Sombroek et al. 1993). Tiessen et al. (1994) calculated a mean turnover rate of 4 years for particulate SOM in undisturbed soils of the Venezuelan rain forest. These phenomena explain why most of the tropical soils become infertile within a few years after cultivation. Soil amelioration with mineral fertilizers or organic manure is often limited by high costs, low cation exchange capacity of the soils, and poor knowledge about the nutrient depletion from the composts (Tiessen et al. 1994).

On the other hand, in the Brazilian Amazon region, patches of sustainable and very fertile anthropogenic dark earths, locally known as *terra preta* (*do índio*) (Smith 1980, 1999; Zech et al. 1990; Sombroek et al. 1993; Denevan 1996; Woods and McCann 1999; Glaser et al. 2001), are embedded in the infertile Ferralsol landscape. The development of *terra preta* sites can be linked to pre-Columbian native populations, which were in many cases large and sedentary, particularly along rivers (Smith 1980; Whitmore 1993; Heckenberger et al. 1999).

Terra preta sites are characterized by higher pH values, higher cation exchange capacities, higher base saturations, and higher nutrient levels compared to the surrounding infertile soils (Zech et al. 1990). The most important difference, however, seems to be the development of a thick (up to 1 m) A horizon (Zech et al. 1990). After forest clearing, SOM within this horizon

[1] Institute of Soil Science and Soil Geography, University of Bayreuth, 95440 Bayreuth, Germany
[2] Institute of Soil Science and Plant Nutrition, Martin Luther University Halle-Wittenberg, 06099 Halle (Saale), Germany

seems to be stable for decades (Smith 1980; Francis and Knowles 2001). The existence of *terra preta* proves that even infertile Ferralsols can be converted to fertile and sustainable soils. While it seems well proven that the formation of the stable humic A horizon is caused by anthropogenic formation of pyrogenic carbon which contributes considerably to the total carbon stock in the dark earths (Glaser et al. 2000a, b, 2001), the origin of the sustainable fertility is still a matter of debate. Therefore, the objective of this study was to investigate to what extent these advantageous soil properties of the dark earths can be related to the pre-Columbian anthropogenic activities associated with the creation of the thick humic A horizon and to what extent this can be explained by indigenous site-specific effects.

11.2
Materials and Methods

11.2.1
Site and Soil Description

The study was carried out in central Amazonia, Brazil, situated within the inner tropical convergence zone (ITC). Mean annual temperature is 26±3 °C and mean annual rainfall is 2,050 mm with a dry season between August and November (Otzen 1992). The whole Amazon Basin is geologically dominated by tertiary sediments of the Alter do Chão formation, which overlies the Precambrian Guyana and Brazilian shields in the north and south, respectively (Otzen 1992). Heavily weathered Ferralsols with a predominance of kaolinite in the clay fraction and an accumulation of iron and aluminum oxides due to desilicification developed. For this study, we investigated *terra preta* sites near Manaus and Santarém with nearby soils for comparison. All investigated sites are located on the *terra firme* plateau, about 100–150 m above sea level with a steep slope to the Amazon River. All soils showed no signs of plaggen epipedons and the sampling sites had negligible slopes. Therefore, colluvial and/or alluvial influences can be rejected and surface run-off and run-on are limited to very heavy rain-storm events. Exact locations of the sampled profiles are described by Glaser et al. (2000a).

Terra preta sites were identified by their thick black A horizons of up to 1 m. Ceramics and charcoal particles were always present throughout the profile. Surrounding Ferralsols sometimes had visible charcoal as well, but only in the upper horizon. For sampling the *terra preta* soils, first the center of the distribution was identified by thoroughly augering the whole site including the surrounding soils. Choosing an adequate control soil which was not influenced by pre-Columbian anthropogenic activity was often difficult as the *terra preta* formations diminish continuously into the surrounding soils. Our criteria for non-*terra preta* sites were (1) the absence of ceramics and (2) the thickness of the A horizon. *Terra preta* and control soil pedons were sampled due to diagnostic horizons, with subsoils being included for a mineralogical comparison of the soil pairs. To investigate textural effects we took samples along a textural gradient from loamy sand to clay. The sandy soils were sampled around Manaus (TP1–4, Fer1–4), whereas the clayey soil pairs (TP5, Fer5) were sampled in the vicinity of Santarém (Glaser et al. 2000a). Although the investigated sites were distributed over a huge area, a comparison among the sites seems appropriate because the pattern of indigenous settlement of the prehistoric population leading to the formation of *terra preta* sites appears to be similar in the whole Amazon region (DeBoer et al. 1996; Denevan 1996). Locations and descriptions of the investigated soils are given by Glaser et al. (2000a).

On the sandy *terra preta* sites near Manaus we identified papaya (*Carica papaya*), passiflora (*Passiflora* sp.), and mango (*Mangifera indica*) plantations, whereas the adjacent Oxisols were dominated by manioc (*Manihot esculenta*) plantations. The clayey soils were covered by secondary rain forest (*capoeira*) with remnants of an abandoned rubber (*Hevea brasiliensis*) plantation.

11.2.2
Methods

Analyses of pore volume and bulk density were conducted according to Schlichting et al. (1995) on undisturbed volume samples (100 ml) which were stored at 4 °C prior to analyses. All other investigations were carried out on air-dried and sieved (<2 mm) soil samples (fine earth). The specific surface area was determined according to the method of Kutilek (Schlichting et al. 1995). pH values were measured in 1 m KCl using a soil suspension with a soil to water ratio of 1:2.5 (w/v) (Schlichting et al. 1995). Exchangeable cations were measured in a 1-M NH_4Oac extract by atom absorption spectrometry (Varian Spectr AA400), and the effective cation exchange capacity (ECEC) was determined by photometric analysis (RFA, Alpkem) of ammonium re-exchanged with 1 m KCl (Schlichting et al. 1995). Base saturation was calculated as the percentage of Na, K, Ca, and Mg from the total exchangeable cations. The textural composition was determined according to Christensen (1992), slightly modified. Briefly, 30 g of soil was dispersed ultrasonically in 150 cm^3 of water using a probe-type disintegrator (Branson Sonifier W450). Coarse sand was separated after an energy input of 60 J cm^{-3} by wet sieving. After an additional energy input of 1,500 J cm^{-3}, fine sand was separated by wet sieving, silt and clay by sedimentation. Coarse sand and fine sand were dried at 40 °C, while silt and clay were freeze dried. Clay minerals were identified according to characteristic d_{001} values obtained with an X-ray diffractometer (XRD, Siemens 5000).

Total organic carbon (TOC), total nitrogen (N), and total sulfur (S) were determined by dry combustion and thermal conductivity detection on an elemental C/N/S analyzer (Elementar, Vario EL).

Pyrogenic carbon was analyzed according to the method of Glaser et al. (1998). Briefly, 0.5 g of ground fine earth was digested with 32 % HCl for 4 h at 170 °C in a high-pressure digestion apparatus. The residue was washed thoroughly with de-ionized water, dried, and oxidized with 65 % HNO_3 for 8 h at 170 °C in a high-pressure digestion apparatus. The residue was washed thoroughly with de-ionized water and the combined filtrate collected and filled up to 10 ml. A 2-ml aliquot was cleaned-up subsequently after addition of citric acid as an internal standard using Dowex 50 WX-8 (200–400 mesh) columns. After drying and derivatization, benzenecarboxylic acids (BCA) were separated by gas-liquid-chromatography and flame ionization detection. The sum of the yields of BCA with three to six carboxylic groups were adopted as a measure for the amount of pyrogenic carbon.

It should be noted that in the meantime it turned out that the HCl treatment may cause artifacts, especially in SOM-rich soils. Thus, in the future HCl should be replaced by 4 M TFA. However, in Mollisols from the North American prairies spanning a wide range of pyrogenic carbon contents, a strong correlation was found for pyrogenic carbon contents obtained with the HCl and a TFA pretreatment (Glaser and Amelung 2003). The soils were classified according to the guidelines of the US Soil Taxonomy (Soil Survey Staff 1998) and the FAO/UNESCO soil map of the world (FAO 1998).

11.2.3
Statistical Analysis

All data were statistically analyzed using Excel 7.0 or Statistica 5.1 for WinNT 4.0. Normal distribution of the data and paired differences were tested using the Levene test. The homogeneity of variances of the normally distributed data was tested with an f-test. Significant differences between the mean values of parameters from *terra preta* and adjacent Ferralsols were tested subsequently using MANOVA followed by the Scheffé test (Hartung et al. 1993; Sokal and Rohlf 1995). The nonparametric Wilcoxon test was applied unless the prerequisites for the t-test were accomplished (Sokal and Rohlf 1995). Factors responsible for the variance of the measured data were extracted using principal component analysis. After varimax rotation of the variables, factors with an eigenvalue greater than 0.7 were considered to be strong and factors between 0.50 and 0.70 to be weak.

11.3
Results and Discussion

11.3.1
Soil Classification

According to Soil Survey Staff (1998), the sandy *terra preta* sites were classified as coarse-loamy, kaolinitic, isohyperthermic Anthropic Humic Xanthic Kandiudox or Anthropic Humic Rhodic Kandiudox, the surrounding soils as coarse-loamy, kaolinitic, isohyperthermic Xanthic Kandiudox or coarse-loamy, kaolinitic, isohyperthermic Rhodic Kandiudox. The clayey *terra preta* and adjacent sites were identified as very fine, kaolinitic, isohyperthermic Anthropic Humic Xanthic Kandiudox and very fine, kaolinitic, isohyperthermic Xanthic Kandiudox, respectively. According to FAO (1998), *terra preta* sites were classified as Fimic Anthrosols (TP1 and 4) and Aric Anthrosols (TP2, 3, and 5). The corresponding native soils were identified as Ferric Acrisols (Fer1–3), Xanthic Ferralsols (Fer4) and Rhodic Ferralsols (Fer5). For simplicity, in the following all anthropogenic dark earths will be referred to as *terra preta* and all adjacent soils as Ferralsols.

11.3.2
Soil Physical Parameters

The texture of the investigated soil pairs followed a gradient with increasing clay and decreasing sand content (Table 11.1). The textural composition of each *terra preta* was comparable to its adjacent Ferralsol, in both the topsoil and subsoil (Table 11.1). This indicates that each soil pair developed from the same parent material and that no new soil material was deposited by human activities. This is corroborated by the fact that also the pattern of clay mineralogy is comparable. In all samples, only kaolinite could be identified as the single clay mineral by XRD-diffractometry (data not shown).

Terra preta soils had better hydrological conditions than the Ferralsols. In the sandy soil pairs, macropores $>50\,\mu m$ covered between 18 and 32% for *terra preta* sites and between 13 and 21% for the Ferralsols. Also, the specific surfaces of *terra preta* soils (TP1 $53.5\,m^2\,g^{-1}$, TP3 $52.5\,m^2\,g^{-1}$, TP5 $140\,m^2\,g^{-1}$) were higher in *terra preta* soils than in the corresponding Ferralsols (Fer1 $16.5\,m^2\,g^{-1}$, Fer3 $40\,m^2\,g^{-1}$). This indicates that *terra preta* soils contain not only more macropores, but also more micropores compared to surrounding soil.

The field capacity in the topsoil horizons averaged 26±1 and 22±1 mm dm^{-1} of *terra preta* and Ferralsols, respectively. These results are in the expected range for sandy soils (Schachtschabel et al. 1997). Similar trends are expected for the clayey soils due to the pseudo-sand structure in clayey Ferralsols. Therefore, plant growth on *terra preta* stands is favored by better water supply and availability of oxygen in comparison to surrounding Ferralsols.

11.3.3
Nutrient Status

Generally, upland Amazonian soils are poor in nutrients due to prolonged leaching and because the weathering front of the geological substrate is too deep to provide nutrients for plants. Compared to the surrounding Ferralsol, the topsoil of *terra preta* is characterized by significantly higher pH values, TOC, N and S levels, and lower C/N ratios (Table 11.1). ECEC and base satura-

Table 11.1. General characteristics of topsoil (0–10 cm) and subsoil (30–40 cm) horizons of the investigated soils in central Amazonia. *ns* Not significant; *, **, and *** significant differences between *terra preta* (*TP*) and Ferralsols (*Fer*) upon pair-wise comparison at the $P > 0.05$, < 0.05, < 0.01, and < 0.001 levels, respectively; *na* data not available

Profile	pH (KCl)	TOC	N	C/N	S	Sand	Silt	Clay
		$(g\,kg^{-1})$			$(g\,kg^{-1})$		$(g\,kg^{-1})$	
Topsoil (0–10 cm)								
TP 1	6.6	37.4	2.90	12.9	0.39	878	59	50
Fer 1	3.6	13.6	0.88	15.5	0.11	850	13	104
TP 2	5.3	34.5	2.93	11.8	0.35	644	51	310
Fer 2	4.0	13.5	0.96	14.2	0.14	750	121	113
TP 3	4.9	31.7	1.86	17.0	na	442	50	483
Fer 3	4.0	13.5	0.96	14.2	0.14	750	121	113
TP 4	4.2	29.5	2.21	13.3	0.35	366	89	508
Fer 4	4.1	26.0	1.56	16.7	0.19	360	73	566
TP 5	6.4	91.2	6.25	14.6	na	98	147	706
Fer 5	3.8	40.4	2.81	14.4	na	27	17	891
	*	*	*	ns	*	ns	ns	ns
Subsoil (30–40 cm)								
TP 1	5.4	9.2	0.45	20.5	0.13	878	22	100
Fer 1	4.2	3.9	0.26	15.1	0.07	829	16	166
TP 2	4.3	10.8	0.74	14.5	0.15	523	20	431
Fer 2	4.0	5.5	0.49	11.3	0.11	682	13	296
TP 3	4.6	25.7	0.99	25.9	na	352	40	534
Fer 3	4.0	5.5	0.49	11.3	0.11	682	13	296
TP 4	4.1	16.5	1.03	16.1	0.21	356	57	592
Fer 4	5.7	3.8	0.4	9.3	0.09	282	68	649
TP 5	5.8	48.5	2.88	16.8	na	87	83	785
Fer 5	3.9	13.3	1.08	12.3	na	28	15	876
	ns	*	*	*	ns	ns	ns	ns

tion (BS) showing a predominance of Ca were significantly higher in *terra preta* sites compared to the surrounding Ferralsols (Table 11.2). According to Schachtschabel et al. (1997), pH, TOC, and ECEC are highly interrelated. To exclude pH effects on CEC, we determined the potential cation exchange capacity (PCEC) at pH 7.0, which also gave significant differences (not shown here) between the topsoil of *terra preta* and Ferralsol. The clay mineralogy, another factor determining the CEC, is predominated by kaolinite in both soils. Due to the low CEC of this clay mineral of 3–15 $cmol_c\,kg^{-1}$ (AG Boden 1994), the differences in CEC must be related to the SOM. Assuming a CEC of SOM of about 200 $cmol_c\,kg^{-1}$ (AG Boden 1994), it is obvious that SOM in *terra preta* soils contributes more to the CEC than the mineral fraction. In particular, as it is known that a considerable part of the SOM in *terra preta* soils consists of aromatic structures from pyrogenic processes (Glaser 1999; Glaser et al. 2000a, b, 2001), it is assumed that this imposingly high CEC is caused by

Table 11.2. Concentrations of exchangeable cations, effective cation exchange capacity (*ECEC*), base saturation (*BS*), and amounts of pyrogenic carbon (C_{pyr}) of topsoil (0–10 cm) and subsoil (30–40 cm) horizons of the investigated *terra preta* soils (*TP*) and surrounding Ferralsols (*Fer*) and statistical differences of pairwise comparison. *ns* Not significant and * significant differences at the $P > 0.05$ and < 0.05 levels, respectively; *b.d.* below detection limit

Profile (mmol$_c$ kg^{-1})	Ca	Mg	K	Na	Fe	Mn	Al+H	ECEC	BS (%)	C_{pyr} (g kg^{-1})
Topsoil (0–10 cm)										
TP 1	85.7	26.3	1.49	0.09	0.54	0.07	b.d.	114.2	99	4.33
Fer 1	b.d.	0.4	0.22	b.d.	1.04	0.01	8.94	10.3	3	0.92
TP 2	58.3	15.9	1.35	b.d.	1.08	0.09	0.79	77.5	97	4.21
Fer 2	0.3	0.3	0.25	0.73	0.50	b.d.	16.60	18.5	8	0.63
TP 3	95.0	26.2	0.92	b.d.	2.30	b.d.	0.98	125	97	6.52
Fer 3	0.3	0.3	0.25	0.73	0.50	b.d.	16.60	18.5	8	0.63
TP 4	18.4	4.7	0.63	0.38	0.42	0.28	8.57	33.4	72	4.96
Fer 4	12.8	5.1	1.57	0.60	0.32	b.d.	5.83	26.3	5	3.26
TP 5	418.3	65.0	1.89	0.79	b.d.	b.d.	7.6	409.3	100	23.92
Fer 5	6.3	3.5	1.20	0.81	b.d.	b.d.	10.8	22.6	52	0.80
	*	*	ns	ns	ns	ns	*	*	*	*
Subsoil (30–40 cm)										
TP 1	25.8	3.3	2.39	b.d.	0.76	0.04	b.d.	32.0	97	21.35
Fer 1	b.d.	0.6	0.04	b.d.	0.74	0.04	6.82	7.9	4	0.11
TP 2	16.2	6.3	0.26	b.d.	0.18	0.03	2.54	25.4	89	2.91
Fer 2	0.2	0.2	0.04	0.50	0.29	b.d.	15.26	16.4	5	0.13
TP 3	65.1	13.0	0.34	0.15	0.31	0.01	0.39	79.3	99	6.04
Fer 3	0.2	0.2	0.04	0.50	0.29	b.d.	15.26	16.4	5	0.13
TP 4	3.2	1.7	0.18	b.d.	0.14	0.14	15.15	20.0	23	5.25
Fer 4	0.2	1.0	0.57	0.46	0.23	b.d.	6.10	8.5	25	0.18
TP 5	154.3	17.1	0.26	0.75	b.d.	b.d.	7.0	289.4	97	8.85
Fer 5	1.9	0.7	0.14	0.93	b.d.	b.d.	6.4	10.1	36	0.10
	*	*	ns	ns	*ns	ns	ns*	*	**	**

partly oxidized pyrogenic carbon forming carboxylic groups on the edges of the aromatic backbone. This assumption is corroborated by the highly significant ($P < 0.001$) correlation between the ECEC and the amounts of pyrogenic carbon (Table 11.2, Fig. 11.1). Moreover, from this relationship it can be calculated that 1 g of pyrogenic carbon contributes 1.78 cmol$_c$ to the ECEC. Compared to the 'average' SOM, the contribution of pyrogenic carbon to the ECEC is by a factor of almost 10 higher, at least in the soils investigated in this study. This factor will increase even more if an overestimation of the absolute amounts of pyrogenic carbon due to the HCl pretreatment is assumed. In both cases, pyrogenic carbon is highly oxidized in *terra preta* soils and it can be concluded that in highly weathered tropical soils, SOM and especially pyrogenic carbon play a key role in maintaining soil fertility.

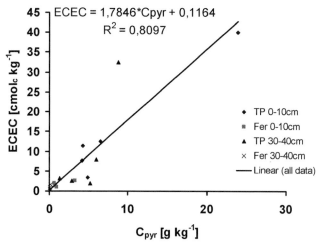

Fig. 11.1. Correlation between the contents of pyrogenic carbon (C_{pyr}) and effective cation exchange capacity (*ECEC*) in *terra preta* (*TP*) and adjacent Ferralsols (*Fer*)

The higher pyrogenic carbon and nutrient levels of *terra preta* sites are certainly due to the traditional land use of the native inhabitants of the Brazilian Amazon region. Repeated superficial slash and burn of a small secondary forest plot, together with ash and incompletely burned residues of smoldering fires for cooking remained on the nutrient-poor soil, where first manioc and later fruit trees could be cultivated (Morawetz 1992). The input of kitchen midden including bone residues of fish, turtles, and game contributed K, P, Ca, and N. Human excrements are considered as important sources of nutrients (especially P and N) in archaeological sites in California (Smith 1980) and they could be a source in *terra preta* as well. However, up to now nothing is known about the extent to which each of these nutrient sources contributed to *terra preta* formation.

To investigate whether soil fertility is due to the in situ formation of *terra preta* and is, hence, caused by human activities and to what extent this is related to textural effects, we conducted a factor analysis using principal component analysis with the variables *terra preta* or Ferralsol (TP or Fer), pH, TOC, N, C/N ratio, S, clay, and ECEC. The results of this analysis have to be interpreted with care due to the low number of observations. However, the variables were varied randomly to check the stability of the multivariate test. After varimax rotation to minimize the number of variables with a high factor load, the program extracted two principal factors (Table 11.3).

To identify the variables explaining the factors, eigenvalues greater than 0.70 are recommended (Sokal and Rohlf 1995). The eigenvalue of a factor is a measure of the amount the factor contributes to the total variance of all variables of the model (Brosius and Brosius 1995). Additionally, we considered variables between 0.50 and 0.70 as weak (Brosius and Brosius 1995). Although there is a large variation in the chemical and physical qualities of *terra preta* sites, related to a different Amerindian influence with respect to

Table 11.3. Results of principal component analysis of soil parameters representing soil fertility measured in topsoil horizons (most 0–10 cm) of *terra preta* and Ferralsol. Factor loads of the analyzed variables and explained variance of extracted principal factors. *Bold* and *italic numbers* indicate strong and weak factor loads, respectively. Both factors are influenced greatly by the anthropogenic impact on the in situ formation of *terra preta* from a Ferralsol; factor 1 includes additionally the texture

Variable	Factor 1	Factor 2
Terra preta or Ferralsol	*0.51*	**0.72**
pH	*0.64*	*0.65*
TOC	**0.98**	0.15
N	**0.93**	0.28
C/N ratio	0.12	**−0.81**
S	0.27	**0.84**
clay	*0.67*	−0.36
ECEC	**0.91**	0.17
Explained variance	58 %	23 %

intensity and duration, some characteristic trends can clearly be discerned. In the topsoil, factor 1 explaining 58 % of the variance of the observed variables was dominated by TOC, N, ECEC, and weakly by clay, pH, and TP or Fer. Factor 2 (23 % of variance) contains the variables S, C/N, and TP or Fer. Therefore, we conclude that factor 1 is dominated by both the texture represented as the variable clay (factor load 0.67) and the anthropogenic impact as the variable TP or Fer (factor load 0.51) influencing the TOC, N, ECEC, and pH. This suggests a synergistic effect between the human impact and texture. Factor 2 is independent of the texture, but it is influenced strongly by the in situ formation of *terra preta*, which is assumed from the high factor load (0.72) of the variable TP or Fer. pH and S contents are independent of the texture. It is interesting to note that the C/N ratio in the topsoil horizon is negatively correlated with the *terra preta* formation, whereas there were no paired differences (Table 11.1). This could be due to an input of charcoal in the topsoil of Ferralsols after burning, but the *terra preta* soils are actually intensively cultivated without burning; they received high amounts of pyrogenic carbon during their habitation of the pre-Columbian Amerindians.

With increasing soil depth, the differences between the nutrient contents of *terra preta* soils and the adjacent Ferralsols diminished. The chemical properties of subsoil horizons (most 30–40 cm) of *terra preta* and Ferralsol were significantly different, but the differences were smaller than in the topsoil horizon (Tables 11.1 and 11.2), indicating that *terra preta* is formed by an in situ process from the top of a Ferralsol. Principal component analysis with the same variables applied to the topsoil again gave only two factors explaining 72 % of the variance (Table 11.4), which could be interpreted again as texture plus human impact (factor 1, 51 %) and the human impact alone (factor 2, 21 %). In this horizon, however, factor 1 has a smaller factor load for the

Table 11.4. Factor loads of soil parameters representing soil fertility measured in subsoil horizons (most 30–40 cm) of *terra preta* (*TP*) soils and adjacent Ferralsols (*Fer*) and explained variance of extracted principal factors. *Bold* and *italic numbers* indicate strong and weak factor loads, respectively. Factor 1 represents the texture and factor 2 the human impact on in situ formation of *terra preta* from a Ferralsol

Variable	Factor 1	Factor 2
TP or Fer	0.34	**0.91**
pH	*0.61*	−0.02
TOC	**0.91**	0.36
N	**0.95**	0.20
C/N ratio	0.15	**0.80**
S	0.04	**0.73**
Clay	*0.67*	−0.13
ECEC	**0.93**	0.17
Explained variance	51%	21%

human impact represented by the variable TP or Fer (topsoil 0.51, subsoil 0.34). The variables explaining 51% of the variance were N, followed by ECEC, TOC, clay, and pH. The factor load of TP or Fer was only 0.34. So we conclude that in the subsoil horizon, the human impact diminished and the high soil fertility (represented by the measured variables) of *terra preta* in this horizon depends primarily on texture, which was indicated by a high factor load of clay (0.67). Factor 2 explaining 21% of the variance is represented again by the human impact (TP or Fer) together with the variables C/N and S. Therefore, we conclude that in the subsoil horizon the differences in variables representing soil fertility depend more on soil texture than on the human impact.

11.3.4
Stocks and Texture Dependence of TOC and N

The TOC and N stocks varied between 147 and 506 and 8 and 33 Mg ha^{-1} m^{-1}, respectively, in the *terra preta* soils and between 72 and 149 and 5 and 10 Mg ha^{-1} m^{-1}, respectively, in the Ferralsols (Fig. 11.2), indicating a highly significant ($P < 0.01$) carbon (factor 1.5–4.6) and nitrogen (factor 1.4–2.3) accumulation in the *terra preta* soils, especially in the agronomically important soil depth between 0 and 30 cm (Fig. 11.2). Sombroek et al. (1993) calculated TOC stocks of 56 Mg TOC ha^{-1} m^{-1} for clayey soils and 34 Mg TOC ha^{-1} m^{-1} for sandy soils as the mean of about 30 profiles of Ferralsols and ferric Acrisols in the Amazon area, which compares well with the carbon stocks of our investigated Ferralsols. Kimble et al. (1990) and De Moraes et al. (1996) calculated TOC stocks of about 100 Mg ha^{-1} m^{-1} for soils under undisturbed vegetation across the Brazilian Amazon Basin based on 1,162 soil profiles of the RADAMBRASIL survey and a digitized Brazilian soil

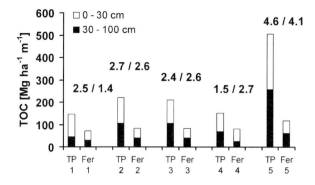

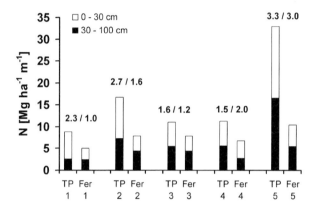

Fig. 11.2. Stocks of a organic carbon (*TOC*) and b nitrogen (*N*) down to 1 m profile depth, divided into the agronomically important 0–30 and 30–100 cm depth intervals in *terra preta* (*TP*) and adjacent Ferralsols (*Fer*). *Numbers above the columns* indicate enrichment of TOC or N between *terra preta* and adjacent Ferralsols in the 0–30 and 30–100 cm depth increments

survey map. This is only one fifth of the TOC stocks found in the *terra preta* soils. Our results show that human impact on infertile tropical soils can result in a tremendous increase in SOM stocks.

On the other hand, the TOC contents also increased with increasing clay content (Fig. 11.3), which is a very common phenomenon in soils (Schachtschabel et al. 1997). This effect is more pronounced in *terra preta* soils than in surrounding Ferralsols, but we found no significant relation between the ratio of TOC in *terra preta* to TOC in Ferralsol and the clay content (topsoil $r=0.5388$ ns, subsoil $r=0.5844$ ns; Fig. 11.3), but partly for N (topsoil $r=0.5388$ ns, subsoil $r=0.8813$, $P <0.05$; Fig. 11.3). These results clearly show that the accumulation of SOM in *terra preta* soils is not primarily a texture effect, but gives credence to a relationship between texture and *terra preta* formation with TOC and N accumulation. This is also emphasized by the principal component analysis (Tables 11.3 and 11.4). Therefore, the general assumption that human impact depletes TOC stocks in soil should be verified again on a broader data basis, although it is known that clay stocks significantly control soil carbon stocks in tropical soils (Zech et al. 1997).

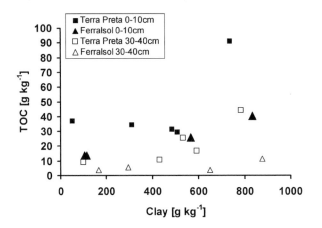

Fig. 11.3. Relation between content of organic carbon (*TOC*) and clay in *terra preta* and adjacent Ferralsols in central Amazonia

11.4
Conclusions

When organic fertilizers such as household refuse, urban waste, agro-processing waste, green manure, forest litter, or sod from external sources are applied to temperate soils, the effect on a long-term basis is obvious. The ancient Plaggen soils of The Netherlands, Flanders, and northern Germany are perhaps the best-known examples. Our results on *terra preta* soils in the Brazilian Amazon region show that, in general, it is also possible in the tropics to increase soil fertility by applying organic and inorganic compounds in spite of higher decomposition rates. Although it is well known that stable organic matter such as pyrogenic carbon plays a key role in maintaining high organic matter levels with a high nutrient retention capacity in *terra preta* soils, high inputs of pyrogenic carbon, e.g. as charcoal or charred residues, alone cannot explain the high nutrient levels of these soils. Although this study gave evidence to the assumption that there was a significant anthropogenic contribution to the nutrient enrichment of *terra preta* soils, at this stage, we can only speculate that in addition to leaving charring residues behind on the soil surface, the aboriginal population enriched and improved the kaolinite-rich soils with organic matter from surrounding terrestrial plants or aquatic grasses and phosphates, nitrogen, and calcium from animals procured through hunting and fishing and by applying excrements, leading to higher trophic levels even after centuries of their abandonment. In the future, efforts should be made to identify and quantify the specific sources of nutrient accumulation with respect to a sustainable agriculture in this area because enhanced biomass production on *terra preta* sites still results in larger carbon and nutrient inputs in the topsoil, resulting in sustainable soils.

Acknowledgements. We kindly acknowledge the financial support of the German Research Foundation, DFG (Gu 406/2-1). We are indebted to Dr. Eugene Balashov for analyzing the specific surface area of selected *terra preta* samples. Dr. Götz Schroth and Dr. Wenceslau Teixeira are acknowledged for logistic support during fieldwork.

References

AG Boden (1994) Bodenkundliche Kartieranleitung, 4th edn. Schweizerbart'sche Verlagsbuchhandlung, Stuttgart

Brosius G, Brosius F (1995) Faktorenanalyse, SPSS. Base system and professional statistics. International Thomson Publishing, Andover, Hampshire

Christensen BT (1992) Physical fractionation of soil and organic matter in primary particle size and density separates. Adv Soil Sci 20:1–90

DeBoer WR, Kintigh K, Rostoker A (1996) Ceramic seriation and settlement reoccupation in lowland South America. Latin Am Antiq 7:263–278

De Moraes JF, Volkoff L, Cerri B, Bernoux C (1996) Soil properties under Amazon forest and changes due to pasture installation in Rondonia, Brazil. Geoderma 70:63–81

Denevan WM (1996) A bluff model of riverine settlement in prehistoric Amazonia. Ann Assoc Am Geogr 86:654–681

FAO (1998) FAO/UNESCO soil map of the world. Revised legend, with corrections and updates. World Soil Resources Report 60, FAO, Rome. Reprinted with updates as Technical Paper 20. IRSIC, Wageningen

Francis JK, Knowles OH (2001) Age of A2 horizon charcoal and forest structure near Porto Trombetas, Para, Brazil. Biotropica 33:385–392

Glaser B (1999) Eigenschaften und Stabilität des Humuskörpers der Indianerschwarzerden Amazoniens. Bayreuth Bodenkund Ber 68:196 pp

Glaser B, Amelung W (2003) Pyrogenic carbon in native grassland soils along a climosequence in North America. Global Biogeochem Cycles 17(2), 1064, doi: 10.1029/2002 GB002019

Glaser B, Haumaier L, Guggenberger G, Zech W (1998) Black carbon in soils: the use of benzenecarboxylic acids as specific markers. Org Geochem 29:811–819

Glaser B, Balashov E, Haumaier L, Guggenberger G, Zech W (2000a) Black carbon in density fractions of anthropogenic soils of the Brazilian Amazon region. Org Geochem 31:669–678

Glaser B, Guggenberger G, Haumaier L, Zech W (2000b) Persistence of soil organic matter in archaeological soils (*terra preta*) of the Brazilian Amazon region. In: Rees RB, Ball B, Campbell C, Watson C (eds) Sustainable management of soil organic matter. CAB International, Wallingford, pp 190–194

Glaser B, Haumaier L, Guggenberger G, Zech W (2001) The *Terra Preta* phenomenon – a model for sustainable agriculture in the humid tropics. Naturwissenschaften 88:37–41

Hartung J, Elpelt B, Klösener KH (1993) Lehr- und Handbuch der angewandten Statistik. Oldenbourg, München, 975 pp

Heckenberger MJ, Petersen JB, Neves EG (1999) Village size and permanence in Amazonia: two archaeological examples from Brazil. Latin Am Antiq 10:353–376

Kimble JM, Eswaran H, Cook T (1990) Organic carbon on a volume basis in tropical and temperate soils. In: Proc Trans 14th Int Congr Soil Science, Kyoto, Japan, 12–18 Aug, Publ 5, pp 248–258

Morawetz W (1992) Die Pflanzenwelt. In: Trupp F (ed) Amazonas. Anton Schroll & Co, München

Otzen H (1992) Amerika vor seiner Entdeckung, Eldorado am Amazonas: Geschichte und Gegenwart einer bedrohten Region. Societäts-Verlag, Frankfurt am Main, pp 31–69

Schachtschabel P, Blume HP, Brümmer G, Hartge KH, Schwertmann U (1997) Lehrbuch der Bodenkunde. Ferdinand Enke Verlag, Stuttgart

Schlichting E, Blume HP, Stahr K (1995) Bodenkundliches Praktikum. Blackwell, Berlin

Smith NJH (1980) Anthrosols and human carrying capacity in Amazonia. Ann Assoc Am Geogr 70:553–566

Smith NJH (1999) The Amazon River forest: a natural history of plants, animals, and people. Oxford University Press, Oxford

Soil Survey Staff (1998) Keys to soil taxonomy. US Department of Agriculture. Pocahontas Press, Blacksburg, Virginia

Sokal RR, Rohlf FJ (1995) Biometry. The principles and practice of statistics in biological research. WH Freeman, New York

Sombroek WG, Nachtergaele FO, Hebel A (1993) Amounts, dynamics and sequestering of carbon in tropical and subtropical soils. Ambio 22:417–426

Tiessen H, Cuevas E, Chacon P (1994) The role of soil organic matter in sustaining soil fertility. Nature 371:783–785

Whitmore TC (1993) Tropische Regenwälder. Eine Einführung. Akademischer Verlag, Heidelberg

Woods WI, McCann JM (1999) The anthropogenic origin and persistence of Amazonian dark earths. Yearbook Conf Latin Am Geogr25:7–14

Zech W (1997) Tropen – Lebensraum der Zukunft. Geog Rundsch 49:11–17

Zech W, Haumaier L, Hempfling R (1990) Ecological aspects of soil organic matter in tropical land use. In: McCarthy P, Clapp CE, Malcolm RL, Bloom RR (eds) Humic substances in soil and crop sciences. Selected readings. American Society of Agronomy and Soil Science Society of America, Madison Wisconsin, pp 187–202

Zech W, Senesi N, Guggenberger G, Kaiser K, Lehmann J, Miano TM, Miltner A, Schroth G (1997) Factors controlling humification and mineralization of soil organic matter in the tropics. Geoderma 79:117–161

12 Use of Space and Formation of *Terra Preta*: The Asurini do Xingu Case Study

FABIOLA ANDRÉA SILVA[1] and LILIAN REBELLATO[1]

12.1
Introduction

In pre-colonial Amazonia as well as presently, the populations that lived and still live in that region have been responsible for enormous alterations in the landscape. This means that the natural environment should not be understood as a determining factor for the way of life of those populations, but as the product of environmental handling carried out by them throughout time (Balée 1994). In this sense, the formation of *terra preta* could also be understood as being one of the results of this cultural creation process of the environment (Petersen et al. 2001). Therefore, a relevant aspect in the study of this material record is to try to understand the cultural processes responsible for its formation. These processes are related to human behavior in subsistence and production activities, specifically the use and disposal of material items.

Contemporary indigenous populations constitute a privileged object of study for the comprehension of these processes; the observation of their daily activities allows us to identify the link that exists between human behavior and the material record. In this chapter, we present results of a first cultural exploration of the indigenous population known as Asurini do Xingu, specifically with reference to the understanding of the way they use their environment in and around the village and the possible consequences of this behavior to the formation of deposits of *terra preta*.

12.2
The Asurini do Xingu and the Use of Space in the Kuatinemu Village

The Asurini do Xingu is a Tupi population whose language belongs to the Tupi-Guarani linguistic family. Since 1985 they have occupied a village located on the right banks of the Xingu River adjacent to the P.I. Kuatinemu, a reservation managed by FUNAI, through its administrative unit located in the town of Altamira, in the State of Pará, Brazil (Fig. 12.1).

[1] Museu de Arqueologia e Etnologia (MAE–USP), Av. Prof. Almeida Prado, 1466, Cidade Universitária Armando de Salles Oliveira, CEP 05508-070, São Paulo, São Paulo, Brasil

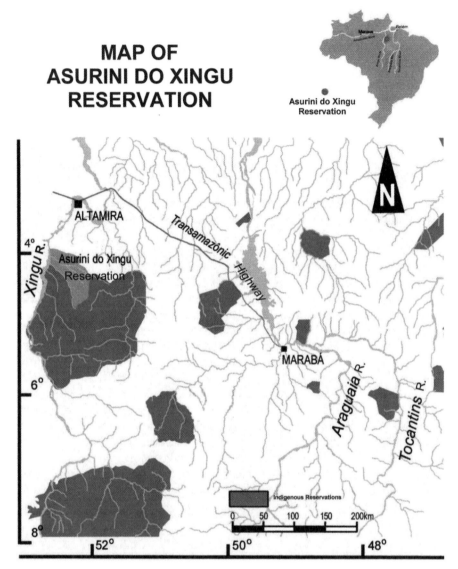

Fig. 12.1. Map of the lower Xingu region in central Amazonia (Brazil) with the Asurini do Xingu reservation area which was used for this preliminary study on recent *terra preta* formation

It is a numerically small population with a contingency of 105 individuals (57 adults and 48 children). Its economy is based on agriculture and hunting, fishing, and gathering of natural resources. All of these activities are performed in cooperation between the sexes. The women are mainly responsible for agriculture and food preparation. The men are involved in hunting, fishing, and food distribution. This relationship between the sexes that orients the subsistence activities is also present in the ritual sphere and in the production of their material culture.

In Asurini society the domestic group is the basic unit of social structure because it is a social and political unit (traditionally identified with the local group) and is responsible for all subsistence activities. Women are the organizers of this social and economic unit, being the basic production unit in Asurini society, while the men's duty is the circulation of the goods produced (Muller 1990).

12.3
Spatial Distribution and Fluxes of Material Used by the Asurini do Xingu in the Kuatinemu Village and Hypothesis on *Terra Preta* Formation

This social unit also defines the spatial distribution of the individuals in the village. The household is positioned in a way to aggregate those belonging to the same domestic group. At the same time the complex of domestic groups is spatially organized in sectors inside the settlement, forming residential groups that maintain reciprocity between each other in terms of distribution and consumption of resources and daily activities.

This same residential structure influences the spatial distribution of the activity areas. These can be divided into public and/or community areas (the ceremonial square and *tavyva* – a communal and ceremonial structure), private and/or domestic areas (domestic units, storing structures, kitchen installations, and yards), and areas of disposal (Fig. 12.2). For a more detailed description of the social environment, see Silva (2000).

The public areas are used for performance of ritual activities and the discussion of political, administrative, social, and economic matters. Characteristically, they show few material remains and are frequently cleaned by the villagers, who deposit the remains in the disposal areas (Figs. 12.3 and 12.4).

The domestic areas are used for the performance of several tasks related to the production, use, and storage of material items, and food preparation and storage (Figs. 12.5 and 12.6). Because they are intensely used, they often show many material remains, but these are also systematically removed to the disposal areas.

The disposal areas are located in the surroundings of the village, normally behind the domestic activity areas. There the remains may be deposited in an intensive (Fig. 12.7) or extensive (Fig. 12.8) manner. In the former, amounts of garbage are formed that might reach – as it was observed in December 2001 – over 1 m in height in an area of about 40 m^2. In the latter, the disposal is more dispersed and less perceptible. In both situations, the presence of vegetation covering the garbage deposits can be observed. Besides, in the domestic area, there might be disposal spots in the form of garbage pits positioned close to the domestic units. In all of these places, various kinds of

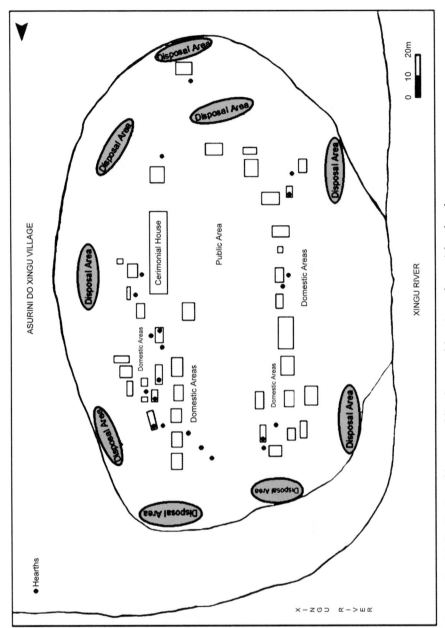

Fig. 12.2. Spatial distribution of houses, public places, and locations of material and garbage deposition within the Asurini do Xingu village

Fig. 12.3. Public area in front of the ceremonial house. This area must be free of material waste, and is cleaned periodically during the year

Fig. 12.4. Public area in front of the ceremonial house. In this picture, a group of women can be seen performing the *'tauva'* ritual, which is associated with war, initiation, and death. Note that the ritual space is free of material remains in comparison to the domestic area in the *lower right* of the photograph

Fig. 12.5. External domestic activity area. This is an area used for the preparation of food and production of materials (ceramics, weapons, adornment, etc.). In this picture, an Asurini woman is shown (Muri) preparing manioc flour in a vessel called *jape'e*

Fig. 12.6. Internal domestic activity area, a place for storing clay, agricultural tools, seeds, and new and used vessels. It is also a place used for the production of ceramic vessels

Fig. 12.7. Area used for extensive disposal of remains on the outskirts of the village. In the *lower left* of the photograph, an area covered by ashes indicates incineration

Fig. 12.8. Area used for intensive disposal of remains, close to a domestic unit. A deposit covered by vegetation can be seen on the *right*, indicating long-term use of the site as a disposal area

remains are deposited, such as food leftovers, traditional material items (pottery, basketry, ropery, etc.), and industrialized products (cans, plastic bottles, fabric, aluminum bowls, etc.). It is important to stress that due to the constant daily depositing of materials throughout 17 years of occupation of the settlement, these areas are characterized by having a great quantity of remains in comparison to the public and domestic activity areas.

This distinct pattern of remains disposal leads us to hypothesize that these are the most favorable places for the formation of deposits of *terra preta*. Another piece of data that reinforces this idea is the Asurini custom to constantly incinerate the deposited remains. We intend to test this hypothesis in a more systematic way in the future with the analysis of soil samples from different activity areas in the village according to the methodology described below.

12.4
Future Study to Investigate Recent *Terra Preta* Formation by the Asurini do Xingu in the Kuatinemu Village

Our research methodology in the Asurini context will be an adaptation of the one developed in August 2002 at the Hatahara archaeological site, located in the town of Iranduba, in the State of Amazonas, Brazil. This site has been investigated since 1999 by the team of *Projeto Amazônia Central* (Central Amazonia Project) under the coordination of Dr. Eduardo Góes Neves. It has an area of approximately 400×400 m and is located at the confluence of the Negro and Solimões Rivers.

In the context of Asurini ethnography, we intend to develop a more detailed topographic map of the settlement than that shown in Fig. 12.2. On the new map, we will plot the different activity areas as well as soil auger tests, following a geographic coordinate plan. The spacing of these auger tests will be proportional to the area of the site, which is around 100×30 m. Therefore, the points for the extraction of samples will be defined according to the configuration of the activity areas so that an adequate sampling of their ethnographic reality may be accomplished. The collection of the soil samples will be in 10-cm-depth increments, having in mind the shorter occupation time of this ethnographic context in comparison to the Hatahara archaeological site. The test pits will be emplaced taking the same criteria into consideration, in this case prioritizing the intensive disposal areas.

As was done at the Hatahara site, we will identify each collected sample with an origin number. The samples will then be analyzed in order to determine the chemical composition of each soil in the different activity areas. Later, these results will be plotted on a topographic map which will allow us to visualize the variability of the soil's chemical composition in relation to the different activity areas.

12.5
Final Considerations

In the studies directed toward an understanding of the processes of *terra preta* formation, up to now few projects have dealt with an anthropological point of view. Nevertheless, considering that this material record is a product of human activity, we must acknowledge that such studies are basic to the understanding of those activities. In this sense, the work we are proposing meets the need for understanding and recording the behavior dynamics that might have been responsible for the genesis and configuration of the archaeologically observed material records.

The Asurini ethnographic context offers one of the multiple explanatory possibilities for the cultural activities that would have led to the formation of this kind of material record. Therefore, anthropological and ethnoarchaeological research on the subject must continue and cultural groups should be studied in order to register the variability of cultural behavior and then develop our interpretations about the *terra preta* formation.

References

Balée W (1994) Footprints of the forest: Ka'apor ethnobotany – the historical ecology of plant utilization by an Amazonian people. Columbia University Press, New York

Muller RP (1990) Os Asurini do Xingu (História e Arte). Editora da UNICAMP, Campinas

Petersen JB, Neves EG, Heckenberger MJ (2001) Gift from the past: *terra preta* and prehistoric Amerindian occupation in Amazonia. In: McEwan C, Barreto C, Neves EG (eds) Unknown Amazon: culture in nature in ancient Brazil. British Museum Press, London, pp 86–105

Silva FA (2000) As Tecnologias e seus Significados. Um Estudo da Cerâmica dos Asurini do Xingu e da Cestaria dos Kayapó-Xikrin sob uma Perspectiva Etnoarqueológica. Tese de Doutorado apresentada no Programa de Pós-Graduação em Antropologia Social

13 Research on Anthropogenic Dark Earth Soils. Could It Be a Solution for Sustainable Agricultural Development in the Amazon?

Beáta E. Madari[1], Wim G. Sombroek[2], and William I. Woods[3]

13.1
Introduction

It has become clear that the economic and cultural growth of humanity is not unlimited, but, to the contrary, it can be seriously constrained by the way resources are administered and managed. The awareness of this perception has made the concept of sustainable development a precondition for economically and socially healthy societies (Hiraoka and Mora 2001). Agricultural systems will have to meet the demands of growing populations, and natural resources will be under increasing pressure (Doran et al. 2002). Changes are happening in our concepts about agricultural development, the use of natural resources, and the stability of the global environment (Doran 2002). The appropriate management of natural resources is unquestionably indispensable in all sectors of agriculture. Governments are investing considerable amounts of resources in research and development projects to provide new products and technologies that may form the basis for sustainable sectoral development.

Today's production systems are often based on the principles and experiences of 'modern' societies which many times were installed in regions with completely different local conditions from those in which these principles formed. This has happened frequently in colonies of European powers and, to some extent, is happening in successor states. In the search for technologies and production systems that may support sustainable development, increasing attention is being given to the knowledge of past native cultures. Doran et al. (1996) have named some basic sustainability strategies for agricultural management systems: (1) conserving soil organic matter (C inputs $\geq$ C outputs); (2) preserving soil structure and reducing soil loss; (3) balancing production and environment through conservation and integrated management systems (optimizing tillage, residue, water, and chemical use) and by synchronizing available nitrogen and phosphorus levels with crop needs during

[1] Embrapa Solos, Rua Jardim Botânico, 1024, 22460-000 Rio de Janeiro, RJ, Brazil
[2] *1934 – †2003
[3] Department of Geography, Southern Illinois University Edwardsville, Edwardsville, Illinois 62026, USA

the year; and (4) relying on more and better use of renewable resources and biodiversity. It is becoming recognized that the production systems of past cultures may provide us with knowledge that could serve as a basis for the development of sustainable management systems in agriculture. Frequently we do not have exact knowledge about the ancient agricultural systems, but by analyzing surviving products, valuable information can be extracted. With this information it may be possible to reconstruct the conditions of its formation and determine the basis of technologies or production systems that will create similar conditions to the original phenomenon.

One such remarkable example is provided by the anthropogenic dark earth (ADE) soils (*terra preta de índio*) of Amazonia. This phenomenon is notable not only for its archaeological and anthropological importance, but also for its most unusual pedological and fertility properties. The soils of Amazonia are generally acidic and have a low nutrient holding capacity, consequently suffer from low fertility, and are clearly a limiting factor for productivity and the sustainability of agricultural systems. However, the ADE soils are media of low input and highly productive local agricultural activities. Because of their peculiar characteristics and their importance in the many local economic structures (German 2003), they have become the focus of attention as a phenomenon that may be able to provide information for the development of economically and ecologically sound, modern production systems that can contribute to the sustainable use of Amazon soils by small-holder farmers, as well as the conservation of the primary forests.

13.2
The Importance of *Terra Preta* Research

The areas of the occurrence of ADE soils are important subjects for cultural heritage preservation (Instituto do Patrimônio Histórico e Artístico Nacional 1988). Archaeological, ethnographical, anthropological, and geographical studies have been long carried out in relation to the ADE soils (e.g., Eden et al. 1984; Herrera et al. 1992; Correa et al. 1994; Mann 2000a, b). Significant soil, geochemical, and pedological information often results from these studies (e.g., Smith 1980; Pabst 1985, 1991; Kern and Kämpf 1989; Costa and Kern 1999; Woods and McCann 2001), but it has not been widely utilized by the agricultural community. Recently, however, research initiatives have been undertaken to explore this 'buried' information for agronomic purposes, especially as bases for conservation agriculture and sustainable development. The anthropogenic dark earth soils have long been known by the pedological society, but they have rarely been included on soil maps because of their individually limited distributions and consequently have generally not been considered to be of economic significance (Woods 1995). In spite of this sentiment, numerous studies have provided reasonable information on the chemical and fertility properties of ADE soils (e.g. Sombroek 1966; Kern and Kämpf 1989; Pabst 1991). Our objective in this chapter is to highlight the importance of research on the ADE soils from an agronomic and agro-ecological point of view.

It is known that ADE soils, compared with the surrounding background soils with no anthropogenic A horizon, exhibit elevated concentrations of plant macronutrients such as calcium (Ca), magnesium (Mg), phosphorus (P), and certain micronutrients, e.g., zinc (Zn), manganese (Mn), and copper (Cu). Other fertility properties like cation exchange capacity (CEC) and base saturation (BS, %) are also high. The aluminum (Al^{3+}) saturation is low and the chemical reaction (pH) of the ADE soils is more favorable than of the surrounding soil profiles. The organic carbon content (OC) of the anthropogenic horizon of ADE soils is also generally higher compared to the OC concentration in non-anthropogenic soils of the region. The most intriguing property of ADE soils is, however, the sustainability of their fertility as witnessed by outside observers and confirmed by local producers. The effect of agricultural use on the fertility of ADE soils, compared with non-anthropogenic soils of the region, is less destructive because of the higher buffer capacity of ADE soils. Within the region, this is a unique characteristic of anthropogenic dark earth soils.

The ADE soils apparently have developed internal microecosystems that do not become impoverished rapidly in tropical conditions to which they are exposed, or under agricultural use. There are various hypotheses about the formation processes of ADE soils. The currently most accepted one suggests that they were formed unintentionally by prehistoric human populations, although substantial evidence exists that intentionality was practiced at many locations (Woods and McCann 2001). Today's archaeological sites were once settlements in the prehistoric past. As a result of these occupations, many materials were concentrated at the sites including organic residues of plant (leaves, different palm leaves, manioc skin, seeds, etc.) and animal (bones, blood, fats, feces, shells, turtle shells, etc.) origin together with great quantities of wood ash and associated burning residues. It is believed that, above all, this considerable amount of organic material contributed to the formation of the high-fertility ADE soils (Woods and McCann 2001). It is important to emphasize that the organic matter of the ADE soils is six times more stable than in adjacent non-anthropogenic soil profiles (Pabst 1991). For these reasons many of the ADE areas are currently used by pioneer farmers and other small-holder producers, who manage to obtain high-level productivity without applying fertilizers or with low amounts of fertilizer use.

Partly because of the above-mentioned properties, the ADE soils can be considered extraordinary examples of an equilibrated coexistence of human beings and nature. There exists a great possibility to unravel the information 'buried' by indigenous populations and incorporate it into modern technologies that provide environmentally safe and sustainable solutions for the necessities of the producers and inhabitants of Amazonia. The sustainable management of soils, besides providing favorable socio-economic conditions, has to be environmentally appropriate. Studies of small local agro-floristic systems which, depending on the location, exist in great variability have shown that it is possible to establish alternative and promising models that promote sustainable development in the humid tropics (Hiraoka 2001). Such models contribute to the preservation of biodiversity, enrich soils, reduce erosion, and in general contribute to ecological preservation.

In the case of the ADE soils, our recent knowledge is sufficient to evaluate and understand the potential and the importance of this phenomenon in order to create alternative fertility management systems for the highly weathered soils of the humid tropical climatic region. However, exact and specific information for the realization of this activity is still missing, principally that concerning the physico-chemical and biological interactions. Besides, it will not be sufficient merely to be able to recreate edaphic conditions similar to those in ADE soils, but we have to have conditions to evaluate potential environmental effects of a technology or soil management system that results in similar conditions to those represented by ADE soils.

13.2.1
What We Know and Do Not Know About the Formation and Properties of ADE Soils

1. There are theories about the formation of Amazonian ADE soils that have to be confirmed (Woods and McCann 2001).
2. There are studies that describe the high and sustainable fertility of the ADE soils (Ranzani et al. 1962; Falesi 1970, 1972; Smith 1980; Kern and Kämpf 1989; Pabst 1991), but studies about nutrient dynamics in ADE soil environments are still missing.
3. There are studies that show that the physico-chemical and molecular properties of the organic matter of these soils play an important role in the formation of the elevated cation exchange capacity, soil structure, and water retention capacity in the ADE soils, and, consequently, in the formation of their sustainable fertility (Zech et al. 1990; Lima et al. 2001, 2002; Glaser et al. 2001). However, knowledge is missing about the carbon dynamics and about the mechanism of carbon sequestration in these soils.
4. There are studies that have demonstrated that the origin of a substantial proportion of the organic matter in the ADE soils is pyrogenic carbon (Black Carbon or charcoal), and that the formation of the highly stable organic matter of these soils is attributed to chemical and biochemical transformations of carbonized residues resulting from natural or induced fire of the plant biomass. In general terms, it is known that by enzymatic activity (Crawford and Gupta 1993; Hofrichter and Fritsche 1997) or by simple chemical oxidation of charcoal (Kumada 1983; Skjemstad et al. 1996; Golchin et al. 1997) high-reactivity compounds can be formed that feature functional groups able to adsorb nutrients and water (Piccolo et al. 1996), and that conserve the polycyclic aromatic structure that gives stability to these compounds. However, in ADE soils, information is missing

about the exact transformation of the pyrogenic carbon, which is originally highly inert, into an organic substance that is highly reactive.

5. There are few studies about the biodiversity of ADE soils. These works suggest that the biodiversity in these soils is higher than in adjacent nonanthropogenic soils (Antony 1997; Antony and van Roy 2002). However, there is no existing detailed and organized information about this biodiversity and about the effect of the management of ADE soils on biodiversity. There is also no information on the specific function of ADE biodiversity in the development of their sustainable fertility.

13.3
Networking Activities in *Terra Preta* Research

Because of the complexity of the phenomenon of the ADE soils, a wide network of cooperation between national and international institutions and professionals of specific areas of knowledge is necessary. The prerequisite for this is efficient networking activity. An important step in the organization of the networking activity in ADE soils research was the formalization of the various project preparatory committees that form the *Terra Preta Nova* (TPN) Group. This occurred during the meeting of the first international workshop specifically devoted to ADE soils and their replication held in Manaus, Brazil, 13–18 July 2002. The participants of the workshop adopted a unifying logo (Fig. 13.1) and confirmed the aims and content of the TPN project (see Sects. 13.3.2 and 13.3.3 for further information on the participating institutions and coordinating committees, respectively). In general, the aims of this project are to contribute to the sustainable use of Amazon soils by small-holder farmers through the creation of conditions similar to those found in ADE soils of the pre-Columbian Indian communities. The establishment of these conditions would also contribute to the carbon stock and sink functions of the Amazon forests and soils.

Fig. 13.1. *Terra Preta Nova* (TPN) Group logo; design developed by Wim Sombroek

13.3.1
Activities of the Multi-Institutional and Multi-Disciplinary Cooperation of the TPN Group

1. Region-wide surveys and local interviews on ADE occurrences, genesis, and related ethnobiodiversity aspects.
2. Anthropological and archaeological site research and directed salvage activities in areas of rapid settlement (such as Apuí, Humaitá, Itaituba, and Novo Progresso).
3. Routine and specialist analysis on sample material aiming to establish the reasons for the high stability and the high cation exchange capacity of the increased soil organic matter concerned.
4. Laboratory and field experiments on the manufacturing of stable and active organic matter, using early indications from the surveys and interviews.
5. Directed agronomic and agroforestry experiments in farmers' fields to replicate soil fertility conditions similar to those in ADE, in conjunction with existing projects on recuperation of degraded Amazonian lands, on secondary forest management, and on agroforestry practices.
6. Extension of an appropriate package of TPN soil management practices to priority areas of integrated environmental management, for example conservation of natural vegetative cover and sustainable small-holder settlement.
7. Awareness promotion on the value of the indigenous patrimony, through state-level educational publications and reference collections on ADE material, etc.
8. Promotion of ADE-oriented land use inside *terras indigenas*, if so desired by the indigenous population groups involved.

13.3.2
Institutions Within the TPN Group

Participating Brazilian institutions:
- Center for Agricultural Research in eastern Amazonia of the Brazilian Agricultural Research Corporation (Embrapa Amazônia Oriental), Belém.
- Center for Agricultural Research in western Amazonia of the Brazilian Agricultural Research Corporation (Embrapa Amazônia Ocidental), Manaus.
- Emílio Goeldi Museum of Pará (MPEG), Belém.
- Federal University of Pará (UFPA), Belém.
- Federal University of Amazonas (UFAM), Manaus.
- Institute of Environmental Protection of Amazonas State (IPAAM).
- National Institute of Amazonian Research (INPA), Manaus.
- National Institute of Historical and Artistic Patrimony (IPHAN), Manaus and Belém.

– National Institute of Indians, Rio de Janeiro.
– National Soil Research Center of the Brazilian Agricultural Research Corporation (Embrapa Solos), Rio de Janeiro.
– University of São Paulo (USP).

International cooperating institutions:

– Cornell University (Cornell Univ.), Ithaca, New York, USA.
– International Soil Reference and Information Centre (ISRIC), Wageningen, The Netherlands.
– Southern Illinois University Edwardsville (SIUE), Illinois, USA.
– University of Bayreuth (Univ. Bayreuth), Germany.

13.3.3
The Coordinating Committees and Homepages of the TPN Group

National committee:

Dirse Clara Kern (MPEG, kern@museu-goeldi.br), Eduardo Góes Neves (USP), Beáta Emöke Madari (Embrapa Solos), and Newton Paulo de Souza Falcão (INPA)

International committee:

William I. Woods (SIUE, wwoods@siue.edu), Bruno Glaser (Univ. Bayreuth), Johannes Lehmann (Cornell Univ.), and Wim G. Sombroek (ISRIC) Homepages of the TPN Group: www.geo.uni-bayreuth.de/bodenkunde and www.museu-goeldi.br/pesquisa/ecologia/tpa

13.4
Approaches to *Terra Preta* Research for Agronomic Applications

The unraveling of the information necessary for the creation of a technology or management system related to the ADE soils of Amazonia is a complex task that demands a comprehensive approach. The technical–scientific questions provided in Section 13.4.1 must be answered in order to fruitfully proceed. These questions indicate that the approach to the investigation of Amazonian ADE soils is necessarily multi- and transdisciplinary. The cooperation of seemingly distant sciences is desirable for a satisfactory characterization of the phenomenon. The disciplines necessarily involved (not necessarily in this order of importance) are anthropology, archeology, biology, botany, ecology, economics, geochemistry, history, pedology, sociology, and soil science.

13.4.1
Fundamental Questions for the TPN Project

1. How were the ADE soils formed? Was this an unintentional or intentional anthropogenic process, or a combination of these two factors?
2. What are the nutrient dynamics in the ADE soils (e.g., P, Ca, K, Mn, Zn)?
3. What are the carbon dynamics of the ADE soils and what is the mechanism of carbon sequestration in them?
4. How can the transformation of the originally inert pyrogenic carbon into a stable and reactive organic substance in the soil be explained? What are the agents of this process?
5. How and to what extent is the biodiversity of the ADE soils different from that of the adjacent non-anthropogenic soils? How does land use affect the biodiversity of ADE soils?
6. What aspects of traditional (practices by local farmers) use of ADE soils help to maintain their fertility and the sustainability of their fertility?

The 2002 Manaus workshop gathered representatives of these various scientific fields together with the aim of creating the research priorities for a comprehensive project called *Terra Preta Nova* (see above). The TPN project should be dedicated to the thorough scientific cognition of the ADE soils and to proposing research imperatives for the use of the 'buried' knowledge in ADE soils for the creation of modern sustainable agricultural production systems in the Amazon. Such a production system would preferentially target local small producers and family farms as a means to improve their social and economical well-being. The priority topics determined at the workshop gave the basics for the elaboration of an international project called '*Terra Preta Nova. Sistemas indígenas de manejo do solo como base para o desenvolvimento de manejo sustentável da fertilidade de solos na Amazônia*' (*Terra Preta Nova.* Indigenous soil management systems as bases for the development of sustainable fertility management of Amazonian soils). The project is led and administered by the National Soil Research Center of the Brazilian Agricultural Research Corporation (*Embrapa Solos*). The participating institutions are listed in Table 13.1. It is worth noting that the interest in ADE soils research is so high that the number of participating institutions in this project is well over the number of those participating in the TPN Group. The 55 research scientists and technicians involved in the project represent fields from molecular biology through pedology and soil fertility to the computer sciences.

The general objective of this international and transdisciplinary project is to seek and organize existing information about the ADE soils and conduct scientific investigation on the chemical, physical, and biological interactions that may contribute to the formation of the unique fertility properties of the ADE soils. The final goal is to be able to offer research imperatives to develop products or technologies as alternatives for the sustainable management of

Table 13.1. Participating institutions in the *Terra Preta Nova* project

Name of institution	Abbreviation	Country
Alterra Research Institute	Alterra	The Netherlands
Cornell University	Cornell Univ.	USA
Embrapa Milho e Sorgo		Brazil
Embrapa Agrobiologia		Brazil
Embrapa Amazônia Ocidental		Brazil
Embrapa Amazônia Oriental		Brazil
Embrapa Instrumentação		Brazil
Embrapa Rondônia		Brazil
Embrapa Solos		Brazil
Faculdade de Ciências Agrárias do Pará	FICAP	Brazil
Instituto Nacional de Pesquisas da Amazônia	INPA	Brazil
International Soil Reference and Information Center	ISRIC	The Netherlands
Museu Paraense Emílio Goeldi	MPEG	Brazil
Plant Research International, Business Unit Crop and Production Ecology	PRI	The Netherlands
Southern Illinois University Edwardsville	SIUE	USA
Universidade de São Paulo, Centro de Energia Nuclear na Agricultura	CENA	Brazil
Universidade Estadual do Pará	UEPA	Brazil
Universidade Federal de Viçosa	UFV	Brazil
Universidade Federal do Amazonas	UFAM	Brazil
Universidade Federal do Pará	UFPA	Brazil
University of Bayreuth	Univ. Bayreuth	Germany
University of Florida	UF	USA

the fertility of Amazonian soils. The specific objectives of the project are listed in Section 13.4.2 below.

13.4.2
Specific Objectives of the TPN Project and Their Justification

1. *Study of the physical properties of ADE soils under different land uses, compared to adjacent soils with no anthropogenic A horizon. Model different scenarios as an exploratory tool to forecast impacts in the hydrological cycle caused by land-use change.*
   The knowledge of the hydrological cycle of the ADE soils under natural conditions and different land use is fundamental not only to understand the way these soils behave, but also to understand their genesis. In recent years mathematical models and simulation techniques have been developed to describe the nature of water in the soil–plant–atmosphere system in a dynamic manner. The modeling of these processes also may help in the evaluation of different scenarios of development in the northern region of Brazil (van Genuchten et al. 1991; Hopmans et al. 2002).

2. *Study of the management of the fertility of ADE soils.*

The evaluation and organization of fertility-related data of ADE soils and the nutritional state of cultivated plants on these soils are a basic move to understand their importance in local production systems. It is important to understand the cycle and availability of plant nutrients, especially phosphorus. The characterization and diagnosis of traditional ADE soil use by local farmers can lead us to identify aspects of land use that may help us to understand how the sustainability of ADE soils fertility is maintained.

3. *Characterization and genesis of the organic matter of ADE soils.*

To appreciate the unique fertility properties of ADE soils, it is necessary to understand the nature and the physico-chemical properties of their organic matter. The evaluation of the effect of agricultural use on the quality and quantity of the organic carbon of the ADE soils in the anthropogenic horizon can provide information on useful and adequate management practices. The identification of the principal mechanisms and reactions of the stabilization of the organic carbon in ADE soils provides information on their capacity to sequester carbon.

4. *Study of the biodiversity of ADE soils.*

One of the virtually unknown aspects of ADE soils are their biological properties. There is practically no available information on the microflora and -fauna or on the composition of vegetation on this particular biotope. Preliminary data have been obtained only recently on the invertebrate community of one ADE site (Antony 1997; Antony and van Roy 2002). It is therefore of great importance to obtain and organize biological (microbial, faunal, and floral) data on the ADE soils. Useful information can be obtained by the identification of the species of the macro-, meso-, and microorganisms; by the determination of the community structure of the invertebrates; by the identification of functional groups; by the evaluation of the potential of the fauna of ADE soils to colonize less fertile soils; by the quantification of carbon and nitrogen levels of the soils' microbial biomass; by the identification of invasive and useful plants in ADE sites; and by the identification of the use of the flora at ADE sites by local populations.

5. *Elaborate initial experiments for the development of subproducts or products as bases for the development of products or technologies as alternatives for the sustainable management of soil fertility of Amazonian soils, using the existing knowledge and the results of the diagnostic-descriptive part of the project.*

There are two preliminary experiments planned to test the practical possibility to recreate the favorable fertility conditions found in ADE soils. One is the evaluation of the addition of residues of the wood industry in the formation of organic matter in Amazonian soils, including the evaluation of the transformations of the organic matter and the role of the biota in it. The other is the evaluation of organic fertilizers of high stability originating from charcoal and by-products of carbonization. In this activity the viability of charcoal production (and of its by-products) will be evaluated in small properties using agri-industrial residues and biomass produced on the property. The charcoal and the by-products of carbonization will be tested as potential sources of stable organic carbon as a tentative to recreate conditions similar to those observed in ADE soils.

6. *Organize the existing and newly obtained knowledge and information into a database, and analyze these data to provide imperatives or recommendations for the development of products or technologies as alternatives for the sustainable management of soil fertility of Amazonian soils.*

13.5
Expected Results of the Cooperation

The possibility offered by the ADE soil phenomenon is in fact great and promising to the development of products, technologies, or soil management systems for tropical agriculture. The hidden knowledge in ADE soils offers possible alternative solutions for the management of low-fertility soils with low production capacity in Amazonia and possibly in acid tropical soils in general. As a result of the comprehensive study of the ADE soils, the elaboration of a document stating research imperatives for the development of products or technologies for the sustainable management of soil fertility of highly weathered acid Amazonian soils is expected. The future technologies or products created based on the knowledge extracted and organized may have the potential to contribute principally to smaller size community and family farming. The expected advantage of such products or technologies would be the intensification of agricultural production, the production of products of higher economical value, the reduction of the risks of production, and possibly the diminution of production costs. This would result in better-quality food supply in the communities or in the family, in higher income possibilities, and finally in the reduction of social disequilibrium.

It is important to emphasize that many of the ADE sites have a reasonable amount of archaeological material which makes these areas important subjects of cultural heritage preservation. The objective of studying this phenomenon by no means can be the exploration of discovered new sites, but the use of the 'buried' information in these soils. This information should be considered as the intellectual property of the indigenous people of Amazonia. For this reason, it would be fortunate if the administration of a project aiming to study and use the knowledge of this phenomenon stayed with a Brazilian national institution like Embrapa (Brazilian Agricultural Research Corporation) which would ensure proper handling of intellectual property rights and even-handed and socially acceptable distribution of the products or technologies.

The benefits and disadvantages of such products and technologies for the Amazon region have to be carefully considered. They can be bases for the intensification of production systems that can improve the quality of life of local populations. On the other hand, these technologies cannot promote, for the same reason, environmental degradation and deforestation. Careful economic studies are necessary to forecast the effects of intensification of agricultural production in settlements and family and community farms as a result of the use of such products and technologies.

Acknowledgements. The authors would like to thank Wenceslau G. Teixeira, Newton P.S. Falcão, Deborah M.B.P. Milori, Lucille M.K. Antony, Dirse C. Kern, and Vinicius M. Benites for their help in the elaboration of the *Terra Preta Nova* project and, through this, indirectly, in the elaboration of this text.

References

Antony LMK (1997) Abundância e distribuição vertical da fauna do solo de ecossistemas amazônicos naturais e modificados. In: MCT-INPA, DFID (ed) Biomassa e Nutrientes Florestais. Projeto Bionte, Manaus, pp 249–255

Antony LMK, van Roy VMA (2002) Invertebrados do solo de *Terra Preta* Arqueológica no Município de Manacapuru, Amazonas, Brasil. In: Resumos do 19 Congresso Brasileiro de Entomologia, Manaus, 16–21 junho, p 179

Correa CMG, Sena CS, Lopes DF, Kern DC, Silveira IM, Furtado L, Gatti M, Cortez R, Peixoto R (1994) O processo de ocupação humana na Amazônia: Considerações e perspectivas. Bol Mus Para Emílio Goeldi Ser Antropol 9(1):3–54

Costa ML, Kern DC (1999) Geochemical signatures of archaeological sites with Black Earth soils in Amazon region. J Geochem Explor 66:369–385

Crawford DL, Gupta RK (1993) Microbial depolymerization of coal. In: Crawford DL (ed) Microbial transformations of low-rank coals. London, pp 65–92

Doran JW (2002) Soil health and global sustainability: translating science into practice. Agric Ecosyst Environ 88:119–127

Doran JW, Sarrantonio M, Liebig M (1996) Soil health and sustainability. In: Sparks DL (ed) Advances in agronomy, vol 56. Academic Press, San Diego, pp 1–54

Doran JW, Stamatiadis SI, Haberern J (2002) Soil health as an indicator of sustainable management. Agric Ecosyst Environ 88:107–110

Eden MJ, Bray W, Herrera L, McEwan C (1984) *Terra preta* soils and their archeological context in the caqueta basin of southeast Colombia. Am Antiq 49:125–140

Falesi IC (1970) Solos de Monte Alegre. IPEAN, Belém, pp 106–111

Falesi IC (1972) O estado atual dos conhecimentos sobre os solos da Amazônia brasileira. In: IPEAN (ed) Zoneamento agrícola da Amazônia. Bol Téc Inst Pesquisa Agropecuária Norte (IPEAN) Belém, 1st reprint, Embrapa CPATU in 1980, pp 33–64

German LA (2003) Historical contingencies in the coevolution of environment and livelihood: contributions to the debate on Amazonian Black Earth. Geoderma 111:307–331

Glaser B, Guggenberger G, Haumaier L, Zech W (2001) Persistence of soil organic matter in archaeological soils (*terra preta*) of the Brazilian Amazon region. In: Rees RM, Ball BC, Campbell CD, Watson CA (eds) Sustainable management of soil organic matter. CAB International, Wallingford, pp 190–194

Golchin A, Baldock JA, Clarke P, Higashi T, Oades JM (1997) The effects of vegetation and burning on the chemical composition of soil organic matter of a volcanic ash soil as shown by ^{13}C NMR spectroscopy. II. Density fractions. Geoderma 76:175–192

Herrera LF, Cavelier L, Rodriguez C, Mora S (1992) The technical transformation of an agricultural system in the Colombian Amazon. World Archeol 24:98–113

Hiraoka M (2001) Puede la agroforestería entregar lo que promete? El caso de Tomé Açu, Pará, Brasil. In: Hiraoka M, Mora S (eds) Desarrollo sostenible en la Amazonía. Mito o realidad? Ediciones Abya-Yala, Quito, Ecuador, pp 85–101

Hiraoka M, Mora S (2001) Introdución. In: Hiraoka M, Mora S (eds) Desarrollo sostenible en la Amazonía. Mito o realidad? Ediciones Abya-Yala, Quito, Ecuador, p 10

Hofrichter M, Fritsche W (1997) Depolymerization of low-rank coal by extracellular enzyme systems. III. In vitro depolymerization of coal humic acids by a crude preparation of manganese peroxidase from the white-rot fungus *Nematoloma frowardii* b19. Appl Microbiol Biotechnol 47:566–571

Hopmans JW, Simunek J, Romano N, Durner D (2002) Simultaneous determination of water transmission and retention properties – inverse methods. In: Dane JH, Topp GC (eds) Methods of soil analysis. Part 1: Physical methods. Soil Science Society of America, Madison

Instituto do Patrimônio Histórico e Artístico Nacional (1988) Coletânea da legislação de proteção ao patrimônio cultural. IPHAN, Manaus, pp 199–225

Kern DC, Kämpf N (1989) Antigos assentamentos indígenas na formação de solos com *terra preta* arqueológica na região de Oriximiná, Pará. Rev Bras Ciênc Solo 13:219–225

Kumada K (1983) Carbonaceous materials as a possible source of soil humus. Soil Sci Plant Nutr 29:383–386

Lima HN, Benites VM, Schaefer CEGR, Mello JWV, Ker JC (2001) Caracterização de ácidos húmicos extraídos de *Terra Preta de Índio*. In: Resumos do 4 Encontro Brasileiro de Substâncias Húmicas, Departamento de Solos, Universidade Federal de Viçosa, pp 155–156

Lima HN, Schaefer CER, Mello JWV, Gilkes RJ, Ker JC (2002) Pedogenesis and pre-Columbian land use of "terra preta anthrosols" ("Indian black earth") of western Amazonia. Geoderma 110:1–17

Mann CC (2000a). Earthmovers of the Amazon. Science 287:786–789

Mann CC (2000b). The good earth: did people improve the Amazon basin? Science 287:788

Pabst E (1985) *Terra Preta do Índio* – Chemische Kennzeichnung und ökologische Bedeutung einer brasilianischen Indianerschwarzerde. Diploma Thesis, University of Bayreuth, Germany

Pabst E (1991) Critérios de distinção entre *Terra Preta* e Latossolo na região de Belterra e os seus significados para a discussão pedogenética. Bol Mus Para Emílio Goeldi Ser Antropol 7(1):5–15

Piccolo A, Pietramellara G, Mbagwu JSC (1996) Effects of coal derived humic substances on water retention and structural stability of Mediterranean soils. Soil Use Manage 12:209–213

Ranzani G, Kinjo T, Freire O (1962) Ocorrências de "plaggen epipedon" no Brasil. In: Bol Téc-Científico Escola Superiore Agric "Liuz de Queroz" (ESALQ), Universidade de São Paulo, no 5. ESALQ, Piracicaba

Skjemstad JO, Clarke P, Taylor JA, Oades JM, Mcclure SG (1996) The chemistry and nature of protected carbon in soil. Aust J Soil Res 34:251–271

Smith NJH (1980) Anthrosols and human carrying capacity in Amazonia. Ann Assoc Am Geogr 70(4):553–566

Sombroek WG (1966) Amazon soils: a reconnaissance of the soils of the Brazilian Amazon Valley. Pudoc, Wageningen

Van Genuchten MT, Leij FJ, Yates SR (1991) The RETC code for quantifying the hydraulic functions of unsaturated soils. ADA, US Environmental Protection Agency

Woods WI (1995) Comments on the black earths of Amazonia. In: Schoolmaster FA (ed) Papers and Proc Applied Geography Conf, vol 18,, Arlington, pp 159–165

Woods WI, McCann JM (2001) El origen y persistencia de las tierras negras de la Amazonía. In: Hiraoka M, Mora W (eds) Desarrollo sostenible en la Amazonía. Mito o realidad? Ediciones Abya-Yala, Quito, Ecuador, pp 23–30

Zech W, Haumaier L, Hempfling R (1990) Ecological aspects of soil organic matter in tropical land use. In: McCarthy P, Clapp CE, Malcolm RL, Bloom PR (eds) Humic substances in soil and crop sciences: selected readings. ASA, SSSA, Madison, pp 187–202

14 Slash and Char: An Alternative to Slash and Burn Practiced in the Amazon Basin

CHRISTOPH STEINER[1,2], WENCESLAU GERALDES TEIXEIRA[2], and WOLFGANG ZECH[1]

14.1
Introduction

The forested area in the tropics continues to decrease. It is a challenge to preserve large areas of tropical forest to counteract the accelerating climate change and loss of biodiversity. The cumulative deforested area (including old clearings and hydroelectric dams) in Amazonia up until 1991 reached 427,000 km^2 or 11% of the 4 million km^2 original forested portion of Brazil's 5 million km^2 legal Amazon region (Fearnside 1997).

Large-scale cattle ranching is mainly responsible for this decline in forest area. However, new settlers advancing along the roads also contribute to deforestation through slash and burn agriculture. In 1990 and 1991, 31% of the clearing was attributable to small farmers (Fearnside 2001).

Slash and burn is an agricultural technique widely practiced in the tropics and is considered to be sustainable when fallow periods of up to 20 years follow 1–3 years of agricultural activity. In many parts of the world, the increasing population size and socio-economic changes including pioneer settlement made slash and burn agriculture unsustainable, leading to soil degradation. In Rondônia, a state in the southwestern corner of the Brazilian Amazon region, intense migration resulted in an increase in the human population at a rate of 15% per year between 1970 and 1980 – a doubling time of less than 5 years. The population of the northern Amazon region increased by 5% per year over the same period (Fearnside 1983). The soil nutrient availability already decreases after one or two cropping seasons. Subsequently, field crops have to be fertilized for optimum production, or fields have to be abandoned and new forests have to be slashed and burned, the common practice.

Soil nutrient and soil organic matter (SOM) contents are generally low in the highly weathered and acid upland soils of central Amazonia, and soil degradation is mainly caused by a loss of SOM as CO_2 into the atmosphere and of nutrients into the subsoil. This process is well known and explains some aspects of the low fertility levels of many soils in the tropics under permanent cropping systems (Zech et al. 1990). In strongly weathered soils of the tropics, SOM plays a major role in soil productivity because it represents the domi-

[1] Institute of Soil Science and Soil Geography, University of Bayreuth, 95440 Bayreuth, Germany
[2] Embrapa Amazônia Ocidental, CP 319, 69011-970 Manaus, Brasil

nant reservoir of plant nutrients such as nitrogen (N), phosphorus (P), and sulfur (S). Generally, SOM contains 95% or more of the N and S, and between 20 and 75% of the P in surface soils (Zech et al. 1997). SOM also influences pH, cation exchange capacity (CEC), anion exchange capacity (AEC), and soil structure. SOM mineralization decreases the total retention capacity of available cations in tropical soils, where SOM is often the major source of negative charge. Maintaining high levels of SOM in tropical soils would be a further step towards sustainability and fertility on tropical agricultural land, thus reducing the pressure on pristine areas.

14.2
Carbon Emissions in Slash and Burn Agriculture

Tropical forests account for between 20 and 25% of the world terrestrial carbon reservoir (Bernoux et al. 2001). Fearnside (1997) calculated net committed emissions of forest burnings in Amazonia. This is calculated as the difference between the carbon stocks in the forest and in the equilibrium replacement landscape. He estimated the above-ground biomass of unlogged forests at 434 Mg ha^{-1}, about half of which is carbon. In most agricultural systems the tendency has been for population pressure to increase, leading to increased use intensity over time and shorter fallow periods, with resulting lower average biomass for the landscape. The net committed emissions for 1990 land-use change in Brazilian Legal Amazonia were 5% of the total global emissions from deforestation and fossil fuel sources (Fearnside 1997). Although most emissions are caused by medium and large ranchers, the emissions of the small farmer population in the Amazon Basin were estimated to be between 34 and 88 million Mg CO_2-equivalent carbon in 1990 (Fearnside 2001).

Charcoal formation during biomass burning is considered the only way that carbon is transferred to long-term pools (Zech et al. 1990; Glaser et al. 1998, 2001, 2002; Fearnside et al. 2001) and can have important effects on atmospheric composition over geological time scales. At a burn of a forest being converted to cattle pasture near Manaus, charcoal represented just 1.7% of the pre-burn biomass. The mean carbon content of charcoal manufactured from primary forest wood in the Manaus region is assumed to be 75% (Fearnside et al. 2001). Soils under tropical forest contain approximately the same amount of carbon as the abundant vegetation above it, being about 3% in the surface horizon and about 0.5% in the subsurface horizons down to 100 cm depth (Sombroek et al. 2000). The soils of Brazilian Amazonia may contain up to 136 Tg of carbon to a depth of 8 m, of which 47 Tg is in the top 1 m. The current rapid conversion of Amazonian forest to agricultural land makes disturbance of this carbon stock potentially important to the global carbon balance and net greenhouse gas emissions. Soil emissions from Amazonian deforestation represent a quantity of carbon approximately 20% as large as Brazil's annual emission from fossil fuels (Fearnside and Barbosa 1998).

14.3
Black Carbon in Soil – *Terra Preta do Índio*

Little attention has hitherto been given to black carbon as an additional source of humic materials. Black carbon is produced by incomplete combustion of biomass, creating various forms such as charcoal, charred plant residues, and soot.

The soils in Brazilian Amazonia are predominantly Oxisols and Ultisols, but in addition a patchily distributed black soil occurs in small areas rarely exceeding 2 ha. This is the so-called *Terra preta do Índio*. Because of the similarity in texture to that of immediately surrounding soils (from more or less sandy to very clayey), and because of the occurrence of pre-Columbian ceramics and charcoal in the upper horizons, these soils are considered to be anthropogenic (Sombroek 1966; Glaser et al. 2001). According to Sombroek (1966), *terra preta* is very fertile, and after clearing of forests the soils are not immediately exhausted as the Oxisols are. *Terra preta* contains significantly more carbon (C), nitrogen (N), calcium (Ca), and phosphorus (P), and the cation exchange capacity (CEC), pH value, and base saturation are significantly higher in *terra preta* soils than in the surrounding Oxisols (Zech et al. 1990; Glaser et al. 2000).

Terra preta soils contain up to 70 times more black carbon than the surrounding soils. Due to its polycyclic aromatic structure, black carbon is chemically and microbially stable and persists in the environment over centuries (Glaser et al. 2001). C^{14} ages of black carbon of 1,000–1,500 years suggest a high stability of this carbon species (Glaser et al. 2000). It is assumed that slow oxidation on the edges of the aromatic backbone of charcoal-forming carboxylic groups is responsible for both the potential of forming organo-mineral complexes and the sustainable increased CEC (Glaser et al. 2001). It can be concluded that in highly weathered tropical soils, SOM and especially black carbon play a key role in maintaining soil fertility.

Black carbon has become an important research subject (Schmidt and Noack 2000) due to its likely importance for the global C cycle (Kuhlbusch and Crutzen 1995). Long-term studies with charcoal applications are needed to evaluate their effects on sustained soil fertility and nutrient dynamics.

14.4
Slash and Char as an Alternative to Slash and Burn

After clearing the land for agricultural production, farmers use the wood for charcoal production. The charcoal is produced in kilns close to the forest edge (Fig. 14.1). The annual charcoal production of Brazil is 3.3 Gg (Gerais 1985). As of 1987, 86% of Brazil's use of charcoal was for industry (i.e., iron and steel) and 14% for residential and commercial purposes. Of wood used for firewood and charcoal in Brazil in 1992, 69% came from plantations, 29% from firewood collection, and 2% from sawmill scraps (Prado 2000).

Fig. 14.1. The charcoal drawing illustrates slash and char in practice. **A** Vegetable planted in charcoal residues; **B** banana planting hole filled with chicken manure, soil, and charcoal; **C** on the sieve table marketable pieces are separated from charcoal dust and small pieces, which are used in agriculture

Only about 85% of the produced charcoal is marketable. On a sieve table the different sizes are sorted out and filled into sacks for selling. Large amounts of broken charcoal pieces and charcoal dust pass through the sieve (Fig. 14.1C). These charcoal residues can be used in agriculture. The percentage of dust could be increased when organic materials other than wood are included in the charring process.

14.5
Alternative Slash and Char in Practice

There is evidence that slash and char is practiced in a wide area of the Amazon Basin. Coomes and Burt (1999) reported from the Peruvian Amazon near Iquitos that for most households, charcoal production is an integral part of their swidden-fallow practices. On field excursions on an unpaved side road leading from the Brazilian highway BR 174, 30 km into the forest, all visited farmers settled on this road practiced slash and char. The agricultural practice of slash and char is also found in the state of Pará south of the city Belém.

Paragominas is known for its timber industry. There the residues from the sawmills are used for charcoal production, and the residues of charcoal production for agriculture. These findings suggest that charcoal residues are frequently used for agriculture where charcoal is produced.

Farmers around Manaus were visited to observe their agricultural practices. There is a high demand for charcoal in the nearby city of Manaus caused by a preference for barbecued meat (*churrasco*). The majority of the farmers use a permanent kiln. In such a kiln about 1,400 kg of charcoal can be produced per filling. A farmer with no access to the nearby city markets sells the charcoal for US$ 0.17 per 2-kg sack. At the main road (BR 174) which goes from Manaus to Presidente Figueiredo, the sacks already have a price of US$ 0.27. In Manaus the same unit is sold for US$ 0.5–0.67 (US$ 1=3 Brazilian *reais*).

Coomes and Burt (1999) investigated charcoal producers near Iquitos in the Peruvian Amazon region and found that the mean kiln producing 945 kg of charcoal requires a labor investment of 26 days. In their study area, charcoal is produced in an earth-mound kiln, which demands more labor than the permanent kiln type used near Manaus. A charcoal producer near the EMBRAPA experimental research station loads his permanent kiln every 12 days. The labor investment is relatively small for the four production stages. Wood is collected and prepared in 1 day by four workers (family members), combustion is supervised for 4 days, charcoal is cooled for 8 days, and than sacked by four workers in 1 day. Cooling and combustion supervision require almost no labor input, allowing time for new wood procurement, charcoal sacking, or agricultural activities. Charcoal producers close to rivers shorten the cooling process by using water. Either 80–100 50-l sacks with a mean weight of 15 kg or 400–500 sacks weighing 3 kg are filled. The producer sells a large sack for US$ 1.00 and a small one for US$ 0.23. On the nearby major road (AM-010), the big sacks are sold for US$ 1.67 and the small ones for US$ 0.50. The monthly income from charcoal production is between US$ 200 and 260, which is three to four times the minimum salary of US$ 66.

Charcoal producers report that the residues are used for their own agriculture and are also picked up by large-scale farmers. These cash crop farms are usually cleared by bulldozer and do not produce their own charcoal. Four types of charcoal use in agriculture were observed:

1. The residues from charcoal production are mainly used as an amendment in planting holes. Mainly bananas (or other fruit trees) are planted in such holes. Typically, the holes are about 30 cm wide and 50 cm deep. These planting holes are filled with chicken manure, charcoal, and soil (Fig. 14.1B).
2. The slash and char farmers produce a kind of charcoal compost. Around the charcoal kilns they dig holes in which the charcoal residues are deposited in layers alternating with organic matter, ashes, and soil. After 1 year of decomposition the farmers use the created material as fertilizer applied on the soil surface. Analyses of such a charcoal-compost show that it has

a high pH value (6.88 in H_2O) and is extremely rich in Ca (2,360.79 ppm), Mg (1,241.2 ppm), and K (521 ppm).

3. Charcoal residues are used for vegetable and herb production in home gardens. These gardens are planted in elevated planters, and the crops are grown about 1.5 m above the ground to avoid damages caused by domestic animals. These planters are filled with soil, charcoal residues, compost, chicken manure, and other forms of organic matter (Fig. 14.1A).

4. Charcoal residues are applied on the soil surface. Farmers report that this maintains soil moisture, especially during the dry season.

In addition, charcoal and charcoal byproducts are used in more technical ways. Coal from geological deposits and from various specialized procedures was successfully used for soil amelioration. Adding charcoal to soil can significantly increase seed germination, plant growth, and crop yields. Similar observations were made after additions of humic acids from coal deposits (Glaser et al. 2002). In the south of Brazil a liquid called *pirolenhoso* is extracted out of the smoke from charcoal production. This technique comes from Japan and the elixir has been used there for centuries to increase crop productivity and quality and to combat diseases and pests in agriculture. So far not much is known about the chemical composition of the product, which consists of more than 200 chemical compounds. For production the gases from the charcoal kiln are captured and channeled in a way to allow the condensation of the vapor. The extract is applied to the soil in a 1:100 dilution with water. In spite of the lack of research, in practice this byproduct of charcoal production has been showing efficiency in controlling nematodes and diseases. On the other hand, used as fertilizer, it increases the vigor and improves the root building, the productivity, and the resistance of the plants, and it increases the sugar content in fruits, which also have accentuated colors and scents (Glass 2001). A growing number of organic farmers have begun the production of *pirolenhoso*. They sell the product for US$ 0.33 l^{-1} and a farmer can produce about 600 l month^{-1}. The market for the extract is very large in Brazil as well as in other countries, mainly in Japan (Glass 2001).

Other charcoal and byproduct uses include the following. A German company invented a product based on coal for the ecological improvement of all types of soils. The product is sold as a soil conditioner and used in organic farming and for restoration of degraded soils, mainly ski slopes (TERRA-TEC, Finning, Germany). Chicken fodder is supplemented with charcoal. Mario Miamoto, the owner of a large battery farm near Manaus (AM-010, km 38), adds 1 % charcoal to the chickens' nutrition in order to assuage the malodorousness of the manure, thus increasing the chickens' appetite. He uses about 2.5 tons of charcoal waste per month. The same is done with cattle fodder to prevent digestion disorders. Osvaldo Sassaki used charcoal successfully for the development of hydroponic systems at the University of Amazonas. Osvaldo Sassaki used charcoal in these experiments as a nutrient-sorbing material (O.K. Sassaki, pers. comm.).

14.6
Advantages of Slash and Char

The advantages of slash and char agriculture as an alternative to slash and burn need to be investigated in more detail, but the following statements can be made already:

1. Charcoal provides income for rural households. Financial income could be used to buy organic fertilizer such as chicken manure, which is cheaper than charcoal and available around Manaus. The residues from charcoal production together with chicken manure can improve the soil's fertility and decrease the amount of leached nutrients (Lehmann et al. 2003).
2. The income from charcoal marketing provides an incentive for longer fallow periods because households practicing slash and char agriculture prefer secondary regrowth of an age between 8 and 12 years. The removal of wood for charcoal production does not diminish agricultural productivity (Coomes and Burt 1999). The mean age of secondary forest cleared for traditional slash and burn agriculture is 5 years. The total biomass of secondary forest derived from farmland at the average age would be $50\,Mg\,ha^{-1}$, including fine litter and other dead above-ground biomass (Fearnside and Guimarães 1996), which is less than half the amount after a 10-year fallow (Table 14.1). Slash and char as an agricultural practice provides increased soil fertility through active improvement by organic matter applications and by longer fallow periods. Additionally, the increased CO_2 reabsorption in longer fallow periods and the charcoal amendments to soil transfer more CO_2 from the atmosphere into biomass and finally into a stable form of SOM.

Table 14.1. Biomass accumulation of secondary forest in Brazilian Amazonia (Fearnside and Guimarães 1996, with permission of Elsevier). The growth rate is highest in young succession stages, creating an incentive for longer fallow periods in slash and char agriculture. Fallows of between 8 and 12 years are sufficiently long for both charcoal production and agricultural cultivation. (Coomes and Burt 1999)

Age	Live biomass $(t\,ha^{-1})$				Root/ shoot ratio	Average growth rate of total biomass since abandonment	Growth rate of total live biomass in interval^{ha-1}
(years)	Wood	Leaves	Roots	Total live		$(t\,ha^{-1}\,year^{-1})$	$(t\,ha^{-1}\,year^{-1})$
5	29.2	4.0	13.8	47.0	0.42	9.4	9.4
10	70.8	6.0	23.1	99.9	0.30	10.0	10.6
20	110.8	10.0	24.2	145.0	0.20	7.3	4.5
30	113.8	9.5	27.7	151.0	0.22	5.0	0.6
80	135.4	8.0	28.5	171.9	0.20	2.1	0.4

3. Charcoal could improve the soil quality by changing soil physical parameters such as bulk density, water retention, and water-holding capacity, a significant advantage for plants, especially in the 4-month dry season.

4. Charcoal amendments seem to have insect-repellent properties. Farmers report that charcoal amendments in banana planting holes keep the widespread pest *Cosmopolites sordidus* (*broca-da-bananeira*) from affecting the plants. This beetle is a common crop pest in all areas of the world where bananas are cultivated (Fancelli 1999).

5. The regeneration of primary forest species is much greater in areas that are not burnt after felling. An unusually high occurrence of primary forest species from the families Lecythidaceae, Bignoniaceae, and Meliaceae was found in an area of secondary growth near Manaus. The area where original forest was cut, but not burned, to obtain wood for charcoal production is unusually rich in young primary forest species. Far less damage is done to the native gene pool when the area is not burned after clearing. This is true not only because of the propensity of many felled trees to regenerate from stump sprouts, but also because seed material is not destroyed (Prance 1975).

6. The CO_2 balance between biosphere and atmosphere as a result of charcoal production is neutral if regrowing wood from plantations or secondary forest is used. The use of charcoal in agriculture would create a carbon sink as a stable soil carbon pool.

7. An indirect advantage of slash and char is that charcoal could also be produced in the wet season, when burning is not possible. Controlled year-round charcoal production would distribute emissions around the year and reduce the high aerosol emissions during the dry season. Artaxo et al. (2002) predicted that the negative effects of burning are not locally restricted. The emitted aerosols reduce solar radiation by about 40% in the critical PAR region, which could lead to an average 3 °C drop in temperature during the burning seasons over regions as large as 3 million km². The reduced temperature and solar radiation seriously affect photosynthesis, and further damage is caused by the phytotoxic gas ozone. Significant ozone concentrations, in the order of 80–100 ppb, were observed in regions far from the burnings. Furthermore, the aerosol emissions could reduce precipitation in some regions by as much as 30%. Artaxo et al. (2002) assumed that as much as half of the Amazonian forest could be affected by secondary pollution. Altogether, these combined effects could reduce the amount of water evaporating from the Amazon's vegetation, affecting weather worldwide.

14.7
Slash and Char Research Activities

Charcoal powder was tested in a randomized multiple block field experiment near Manaus, Brazil. Charcoal amendments (11 Mg ha^{-1}) elevated the above-ground biomass production significantly on fertilized plots. In the second cropping period the yield of sorghum (*Sorghum bicolor*) was increased by 880 % in comparison with plots receiving just mineral fertilizer without charcoal amendments. Charcoal amendments alone did not increase crop productivity. These results strengthen the hypothesis that charcoal retains nutrients and makes them available.

The claims that charcoal amendments in banana planting holes keep the widespread pest *Cosmopolites sordidus* (*broca-da-bananeira*) from attacking the plants could not be confirmed. In a greenhouse experiment 20 banana plants of two different varieties (Caipira and Prata Zulu) where planted in pots. The soil was amended with chicken manure and lime. Ten plants received additional charcoal amendments (one third of the volume). Four of the ten bananas were infected by *Cosmobolites sordius* in the charcoal treatment, showing clearly that charcoal does not repel those species. On the other hand, five of the ten banana plants in the treatments without charcoal died, apparently because of a lack of drainage. Insufficient water drainage affected mainly the Caipira variety (four of five plants). Insufficient drainage was also observed in a banana plantation north of Manaus (BR 174, km 102) where rotten banana rhizomes were found in planting holes full of standing water.

Greenhouse experiments of Lehmann et al. (2003) showed that charcoal additions increased biomass production of a rice crop by 17 % in comparison to a control on a Xanthic Ferralsol. Combined application of N with charcoal resulted in a higher N uptake than what would have been expected from fertilizer or charcoal applications alone. The reason is a higher nutrient retention of applied ammonium by the charcoal-amended soils.

14.8
Conclusions

The observed effects of charcoal applications in slash and char agriculture seem to match the properties of the fertile anthropogenic *terra preta* soils in the Amazon Basin. Charcoal production is a lucrative activity and transfers SOM into stable pools when residues are used in agriculture. Where charcoal is produced the residues are used for soil amelioration. Farmers evolved various techniques to use charcoal residues. Due to the relatively low nutrient content, charcoal is mixed with chicken manure for planting holes or a nutrient-rich charcoal compost is produced for surface application. This compost could act as a slow-release fertilizer. In our experiments, soil charcoal amendments improved crop growth and yield significantly. We are conducting further experiments to determine the mechanisms of soil improvement through charcoal amendments and the efficiency of slash and char agriculture. Should slash and char become common throughout the tropics it could serve as a significant carbon sink and could improve the sustainability of tropical agriculture.

Acknowledgements. The research was conducted within SHIFT ENV 45, a German–Brazilian cooperation, and financed by BMBF, Germany, and CNPq, Brazil. A financial contribution was given by the doctoral scholarship program of the Austrian Academy of Sciences. We are grateful for Johannes Lehmann's (Cornell University) and Bruno Glaser's (University of Bayreuth) valuable advice and for the fieldworkers' help particularly Luciana Ferreira da Silva and Franzisco Aragão Simão and the laboratory technician, Marcia Pereira de Almeida. We thank Ilse Ackerman for her comments on a draft of this paper.

References

Artaxo P, Martins JV, Yamasoe MA, Procópio AS, Pauliquevis TM, Andreae MO, Guyon P, Gatti LV, Leal AMC (2002) Physical and chemical properties of aerosols in the wet and dry seasons in Rondônia, Amazonia. J Geophys Res 107(Spec Issue 0):14

Bernoux M, Graça PMA, Cerri CC, Fearnside PM, Feigl BJ, Piccolo MC (2001) Carbon storage in biomass and soils. In: Richey JE (ed) The biogeochemistry of the Amazon Basin. Oxford University Press, New York, pp 165–184

Coomes OT, Burt GJ (1999) Peasant charcoal production in the Peruvian Amazon: rainforest use and economic reliance. For Ecol Manage 140:39–50

Fancelli M (1999) Pragas. In: Alves ÉJ (ed) A cultura da banana, aspectos técnicos, socioeconômicos e agroindustriais. Embrapa/SPI&Cruz das Almas. Embrapa-CNPMF, Brasilia, pp 409–431

Fearnside PM (1983) Land-use trends in the Brazilian region as factors in accelerating deforestation. Environ Cons 10:141–148

Fearnside PM (1997) Greenhouse gases from deforestation in Brazilian Amazonia: net committed emissions. Clim Change 35:321–360

Fearnside PM (2001) Effects of land use and forest management on the carbon cycle in the Brazilian Amazon. J Sustain For 12:79–97

Fearnside PM, Barbosa RI (1998) Soil carbon changes from conversion of forest to pasture in Brazilian Amazonia. For Ecol Manage 108:147–166

Fearnside PM, Guimarães WM (1996) Carbon uptake by secondary forest in Brazilian Amazonia. For Ecol Manage 80:35–45

Fearnside PM, Lima PM, Graça A, Rodrigues FJA (2001) Burning of Amazonian rainforest: burning efficiency and charcoal formation in forest cleared for cattle pasture near Manaus, Brazil. For Ecol Manage 146:115–128

Gerais FCTDM (1985) Recuperacão de alcatrão vegetal. STI/CIT, Brasilia

Glaser B, Haumaier L, Guggenberger G, Zech W (1998) Black carbon in soils: the use of benzenecarboxylic acids as specific markers. Org Geochem 29:811–819

Glaser B, Balashov E, Haumaier L, Guggenberger G, Zech W (2000) Black carbon in density fractions of anthropogenic soils of the Brazilian Amazon region. Org Geochem 31:669–678

Glaser B, Haumaier L, Guggenberger G, Zech W (2001) The "*terra preta*" phenomenon: a model for sustainable agriculture in the humid tropics. Naturwissenschaften 88:37–41

Glaser B, Lehmann J, Zech W (2002) Ameliorating physical and chemical properties of highly weathered soils in the tropics with charcoal – a review. Biol Fertil Soils 35:219–230

Glass V (2001) Reportagens Tecnologia-Onde há fumaça há lucro. Globo Rural 188

Kuhlbusch TAJ, Crutzen PJ (1995) Toward a global estimate of black carbon in residues of vegetation fires representing a sink of atmospheric CO_2 and a source of CO_2. Global Biogeochem Cycles 9:491–501

Lehmann J, da Silva Jr JP, Steiner C, Nehls T, Glaser B, Zech W (2003) Nutrient availability and leaching in an archaeological Anthrosol and a Ferralsol of the central Amazon basin: fertilizer, manure and charcoal amendments. Plant Soil 249:343–357

Prado M (2000) The environmental and social impacts of wood charcoal in Brazil. Os Carvoeiros: the charcoal people of Brazil. Wild Images, Rio de Janeiro, 192 pp

Prance GT (1975) The history of the INPA capoeira based on ecological studies of Lecythidaceae. Acta Amazonica 5:261–263

Schmidt MWI, Noack AG (2000) Black carbon in soils and sediments: analysis, distribution, implications and current challenges. Global Biogeochem 14:777–793

Sombroek WG (1966) Amazon soils, a reconnaissance of the soils of the Brazilian Amazon region. Diss, Pudoc, Wageningen, The Netherlands

Sombroek WG, Fearnside PM, Cravo M (2000) Geographic assessment of carbon stored in Amazonian terrestrial ecosystems and their soils in particular. In: Sewart BA (ed) Global climate change and tropical ecosystems. CRC Press, Boca Raton, 483 pp

Zech W, Haumaier L, Hempfling R (1990) Ecological aspects of soil organic matter in the tropical land use. Humic substances in soil and crop sciences; selected readings. American Society of Agronomy and Soil Science Society of America, Madison, pp 187–202

Zech W, Senesi N, Guggenberger G, Kaiser K, Lehmann J, Miano TM, Miltner A, Schroth G (1997) Factors controlling humification and mineralization of soil organic matter in the tropics. Geoderma 79:117–161

Microbial Response to Charcoal Amendments of Highly Weathered Soils and Amazonian Dark Earths in Central Amazonia – Preliminary Results

Christoph Steiner[1], Wenceslau G. Teixeira[2], Johannes Lehmann[3], and Wolfgang Zech[1]

15.1
Introduction

The abundance of charcoal and highly aromatic humic substances in Amazonian Dark Earths (ADE) suggests that residues of incomplete combustion of organic material (black carbon, pyrogenic carbon, charcoal) are a key factor for the persistence of soil organic matter (SOM) in ADE soils which contain up to 70 times more black carbon than the surrounding soils (Glaser et al. 2001). ^{13}C-NMR studies showed that the only chemical structures that appear to survive decomposition processes are mostly due to finely divided charcoal (Skjemstad 2001). Generally, in highly weathered tropical soils, SOM and especially charcoal play a key role in maintaining soil fertility (Glaser et al. 2001, 2002).

Black carbon in soil has become an important research subject (Schmidt and Noack 2000) due to its likely importance for the global carbon (C) cycle (Kuhlbusch and Crutzen 1995). However, charred organic matter (OM) and black carbon in terrestrial soils have been rarely evaluated regarding their importance for nutrient supply and retention (Lehmann et al. 2003).

Addition of charcoal to soil was shown to affect various microbial processes in soil. For example, charcoal stimulates the colonization of crops by indigenous arbuscular mycorrhizal fungi (AMF). ^{13}C NMR and FTIR showed that charcoal is a microporous solid composed primarily of elemental (aromatic) C and secondarily of carboxyl and phenolic C (Braida et al. 2003). Non-local density functional theory (N_2, Ar) and Monte Carlo (CO_2) calculations revealed a porosity of 0.15 cm^3/g, specific surface area of 400 m^2/g, and appreciable porosity in ultramicropores of <10 Å (Braida et al. 2003). AMF can easily extend their extraradical hyphae into charcoal buried in soil and sporulate in the porous particles (Saito and Marumoto 2002). Those pores may offer a microhabitat to the AMF, which can obtain nutrients through mycelia extended from roots (Nishio 1996). The stimulation of AMF due to charcoal is relatively well known, although in practice limited due to its high cost (Nishio 1996; Saito and Marumoto 2002). In a slash and char agricultural

---

[1] Institute of Soil Science, University of Bayreuth, 95440 Bayreuth, Germany
[2] Embrapa Amazônia Ocidental, 69011-970 Manaus, Brazil
[3] Department of Crop and Soil Sciences, Cornell University, Ithaca, New York 14853, USA

practice as an alternative to slash and burn (Lehmann et al. 2002), the costs of charcoal application could be low enough to be profitable (Steiner et al., Chap. 14, this Vol.). Nishio (1996) showed that charcoal can improve nodule formation, nodule weight (2.3 times), and nitrogen uptake (2.8–4 times). Charcoal particles have a large number of continuous pores. Pietikainen et al. (2000) found that charcoal itself supported a microbial community which was small but more active than that of humus. Microcosm studies showed that bacterial growth rate (k) and basal respiration (BR) were higher in charcoal treatments compared to treatments without charcoal. Charcoal has the capacity to adsorb plant-growth-inhibiting organic compounds and may form a new habitat for microbes, which decompose the absorbed compounds (Pietikainen et al. 2000). In the presence of charcoal, seedling shoot and root ratios and nitrogen uptake of a tree species (*Betula pendula*) were enhanced (6.22 times) but only if humus phenolics produced by ericaceous vegetation was present in a boreal forest ecosystem. Fern prothalli were entirely absent in ericaceous substrate unless charcoal was also present. Charcoal maintains a high sorptive capacity for about a century after wildfire, and this sorption is potentially capable of reversing the negative effects of plant species that produce high levels of phenolics (Wardle et al. 1998). Fischer and Bienkowski (1999) and Uvarov (2000) investigated respiration of soil community and decomposition rate of SOM after long-term exposure to smoke emissions from charcoal production in Poland. They found that soil systems in the neighborhood of charcoal kilns have generally a higher level of biological activity. The average density of seedlings was significantly higher in the contaminated soils in comparison with the soils in uncontaminated forest (Uvarov 2000). These examples illustrate the important effects of charcoal on soil ecological processes.

Microorganisms transform and recycle OM and plant nutrients in the soil and are sinks (during immobilization) and sources (during mineralization) of labile nutrients (Stenström et al. 1998). Immobilization could be important as a nutrient retention mechanism in those soils highly affected by leaching. On the other hand, phosphorus (P) is strongly bound to aluminum and iron oxides and is thus not available for plants. Heterophobic phosphate-solubilizing microorganisms make mineral-bound phosphate available by the excretion of chelating organic acids. Microorganisms need large amounts of organic matter before they can excrete organic acids (Nishio 1996). The phosphate and other plant nutrients become available after the OM becomes exhausted and microbial biomass decreases, releasing nutrients into the soil. Kimura and Nishio (1989) showed that insoluble phosphates that are not crystallized can be solubilized by indigenous microorganisms when abundant carbon sources are supplied.

Mainly Japanese farmers utilize microorganisms in the hope of increasing the yield or quality of their crops at a relatively low cost (Nishio 1996). AMF inocula are useful to reduce phosphate fertilizer application.

The microbial biomass can serve as an indicator of the effect of charcoal applications as microbial biomass was shown to react significantly to different management practices (Stenström et al. 1998). Substrate-induced respiration (SIR) also serves to assess the potential performance of microorganisms and this provides hints for nitrogen supply potential, and availability of organic compounds or inhibiting agents like toxic materials (Beck and Bengel 1992).

Therefore, microbial respiration was measured to gain information about the changes in the fertility and quality of tropical soil amended with organic and mineral fertilizers over a 2-year period. We investigated the long-term effects of OM additions in the form of particulate charcoal with respect to its sustainability and resilience against microbial decomposition by investigating ADE. Most models dealing with the turnover of soil organic carbon include a soil carbon pool that either is inert or has a turnover time measured in centuries or millennia (Goldammer 1993; Goudriaan ; Kuhlbusch and Crutzen 1995; Kuhlbusch et al. 1996; Skjemstad et al. 1996; Trumbore et al. 1996; Schmidt and Noack 2000; Verburg et al. 1995). The strongest evidence of the presence of such a pool comes from radio carbon dating of soils that shows that in some cases, even in surface horizons, the mean resistance time for organic carbon exceeds 1,000 years. A strong interaction with the mineral matrix, chemical recalcitrance imparted through strong humification processes, and the presence of significant quantities of finely divided charcoal are the main mechanisms for imparting SOM resistance (Skjemstad 2001). Finely divided charcoal was applied in the experimental field and was studied in respect to microbial respiration in comparison to other organic amendments. Large quantities of charcoal are found in ADE, which is the key for the recalcitrance of the OM (Glaser et al. 2000). Additionally, the suitability of this method to investigate properties of tropical soils under different management was of interest.

15.2
Material and Methods

This study was conducted at the Embrapa (Empresa Brasileira de Pesqusa Agropecuária) Amazônia Ocidental station, near Manaus, Brazil. The average precipitation is 2,503 mm year^{-1} (1971–1993) with a maximum between December and May. The natural vegetation is tropical rainforest. The soils are classified as Xanthic Ferralsols (FAO 1990) and are clayey (with over 80% clay) and strongly aggregated.

15.2.1
Study Design

OM amendments to soil (charcoal powder C_C, chicken manure C_M, compost C_O, and litter L) were tested in a field experiment with respect to their effect on microbial respiration of soil. Respiration curves of top soil samples (0–10 cm) were determined in 13 treatments comprising different combina-

Treatments:

1. Control (C)
2. Litter (L)
3. Burned litter (L_B)
4. Control + mineral fertilizer (F)
5. Chicken manure (C_M)
6. Compost (C_O)
7. Charcoal in powder (C_C)
8. Compost + mineral fertilizer (C_O + F)
9. Charcoal in powder (25%) + mineral fertilizer (C_C+F)
10. Charcoal in powder (C_C½) + compost (C_O½)
11. Charcoal in powder (C_C½) + compost (C_O½) + mineral fertilizer (F)
12. Charcoal in powder (C_C) + compost (C_O½)
13. Charcoal in powder (C_C) + compost (C_O½) + mineral fertilizer (F)

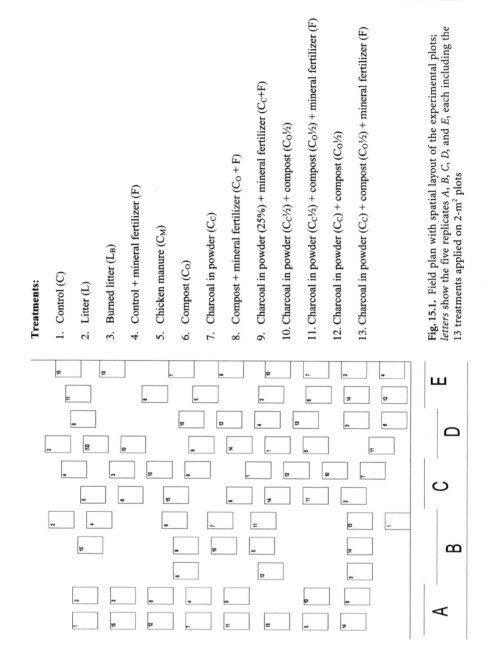

Fig. 15.1. Field plan with spatial layout of the experimental plots; *letters* show the five replicates *A, B, C, D,* and, *E,* each including the 13 treatments applied on 2-m² plots

tions of organic amendments in August 2001 and November 2002. The experiment was laid out in a randomized complete block design using five replicates (Fig. 15.1).

Vegetation litter and root material were removed from the entire field area (45×35 m) and aluminum sheets were used as erosion controls between the plots. The treatments were applied on 4-m² plots (2×2 m) and the amount of applied OM was calculated from the total C content of the materials to increase total soil C content in the 0- to 10-cm depth by 25% (Fig. 15.1).

Table 15.1. Amounts of additions of organic matter (February 2001) and mineral fertilizer (March 2001 and April 2002) to a Xanthic Ferralsol in central Amazonia

Treatment	Organic matter type February 2001 (kg ha^{-1})	Mineral fertilizer type March 2001 (kg ha^{-1})	Mineral fertilizer type April 2002 (kg ha^{-1})
C	None	None	None
L	Litter (13,000)	None	N (55), P (40), K (50), lime (2,800), Zn (7), B (1.4), Cu (0.6), Fe (2.3), Mn (1.6), Mo (0.08)
L$_B$	Burned litter (13,000 dry litter before burning)	None	N (55), P (40), K (50), lime (2,800)
F	None	N (30), P (35), K (50), lime (2,100)	N (55), P (40), K (50), lime (430)
C$_M$	Chicken manure (47,000)	None	None
C$_O$	Compost (67,000)	None	None
C$_C$	Charcoal (11,000)	None	None
C$_O$+F	Compost (67,000)	N (30), P (35), K (50), lime (2,100)	N (55), P (40), K (50), lime (430)
C$_C$+F	Charcoal (11,000)	N (30), P (35), K (50), lime (2,100)	N (55), P (40), K (50), lime (430)
C$_C$^{1}/$_2$+C$_O$^{1}/$_2$	Charcoal (5,500), compost (33,500)	None	None
C$_C$^{1}/$_2$+C$_O$^{1}/$_2$+F	Charcoal (5,500), compost (33,500)	N (30), P (35), K (50), lime (2,100)	N (55), P (40), K (50), lime (430)
C$_C$+C$_O$^{1}/$_2$	Charcoal (11,000), compost (33,500)	None	None
C$_C$+C$_O$^{1}/$_2$+F	Charcoal (11,000), compost (33,500)	N (30), P (35), K (50), lime (2,100)	N (55), P (40), K (50), lime (430)

Mineral fertilizer (NPK and lime) was applied as recommended by Embrapa (Fageria 1998). Organic materials were applied just once at the beginning of the experiment (3 February 2001). Mineral fertilizer was applied in March 2001 and after the second harvest in April 2002 (Table 15.1). Crop residues were not removed from the plots. As a first crop, rice (Oryza sativa L.) was planted followed by three repeated sorghum (Sorghum bicolor L. Moench) crops. Sole compost amendments (C_O) was only studied in the second year due to a broken chamber in the measuring device.

15.2.2
Soil Sampling and Analyses

Soil samples (0–10 cm) were taken after the first harvest in August 2001 and after the fourth harvest in November 2002. In addition, 29 soil samples from 5 different archaeological sites covered by ADE (locally known as terra preta de indio) [Embrapa – Estaçao Experimental do Caldeirão – Iranduba ($n=14$), Fazenda Jiquitaia – Rio Preto da Eva ($n=4$), Lago da Valéria – Parintins ($n=4$), Ramal Acutuba ($n=9$), TP Ramal das Laranjeiras ($n=2$)], plus 5 primary forest and 5 secondary forest sites were sampled and analyzed. For basic soil characterization, we analyzed the soil samples for nitrogen (N), phosphorus (P), magnesium (Mg), calcium (Ca), and potassium (K) and measured the pH. For the extraction of exchangeable P, K, Ca, and Mg, the Mehlich-3 extraction method was used without modification (Mehlich 1984). The filtered solutions were analyzed for P, K, Ca, and Mg using atomic absorption spectrometry (AA-400, Varian Associates, Inc., Palo Alto, California). Total C and N were analyzed by dry combustion with an automatic C/N Analyzer (Elementar, Hanau, Germany). The soil pH was determined in water and KCl using an electronic pH meter with a glass electrode (WTW pH 330). Deionized water and 1 M KCl were applied in the ratio 1:5 soil:dilution medium.

The soil was sieved (<4 mm), humidity was measured, and the samples were stored overnight in darkness at 20 °C. Microbial biomass was determined using basal and substrate-induced respiration (SIR). Respiration measurements were performed with the IRGA-based (infra-red gas analysis) ECT-Soil Respiration Device (ECT Oekotoxikologie GmbH, Germany) according to the procedure described by Anderson and Domsch (1878) and Förster and Farias (2000). The respiration of soil samples (40 g dry weight) was determined by measuring the carbon dioxide (CO_2) production over time in a continuous flow system at a constant flow rate of 300 ml fresh air min^{-1}. A portable computerized photosynthesis measuring system HCM-1000 (Heinz Walz GmbH, Effeltrich, Germany) was used for CO_2 measurement. The central unit of the system consists of an IRGA, a peristaltic air pump, and a mass flow meter that is connected to a measuring chamber (cuvette). It operates in an open flow mode (differential mode), measuring the difference between the CO_2 concentration of the ambient air before and after passing the cuvette. The system is controlled via a computer. To measure soil respiration the central unit was connected to a specially designed rag containing 17 cuvettes. Each cuvette was connected via tubing and solenoid valves to the central unit.

Microbial respiration was measured for 12 h. After glucose addition the measurement continued for an additional 24 h (SIR). Each soil sample was measured once within 1 h over a period of up to 50 h. The SIR method is a physiological method for the measurement of the soil microbial biomass. When easily degradable substrates, such as glucose, are added to a soil, an immediate increase of the respiration rate is obtained, the size of which is assumed to be proportional to the size of the microbial biomass (Stenström et al. 1998). The basal respiration is measured before the addition of the substrate and the SIR shortly after the substrate (240 mg glucose) addition. Microbial respiration was calculated according to:

$$\text{Respiration [nL CO}_2 \text{ min}^{-1} \text{ g}^{-1} \text{ soil]} = (C^*F)/S \tag{1}$$

where C is the IRGA-measured CO_2 value (ppm), F is flow rate through the cuvette (ml min^{-1}), and S is soil dry weight (g).

Microbial biomass was calculated according to Anderson and Domsch (1978):

$$\text{Microbial biomass (}C_{mic} \text{ [}\mu\text{g } C_{mic} \text{ g}^{-1} \text{ soil]} = (R^*40.04)+0.37 \tag{2}$$

where R is respiration (μl CO_2 g^{-1} h^{-1}).

The specific respiration increment was quantified as the slope of the exponential respiration increase after substrate addition when the respiration rate is plotted on a scale against time. This slope was described by:

$$M = N_0 e^{kt} \tag{3}$$

where N_0 is the initial concentration of microorganisms, k is the specific growth rate, and t is time.

The following parameters served as indicators of soil quality, OM turnover, and nutrient availability (Fig. 15.2): basal respiration (BR), substrate-induced respiration (SIR), velocity of population increase (k) after substrate addition (nutrient availability and soil quality), activation quotient (QR=BR/SIR, microbial efficiency), C_{mic}/C_{org} (population density independent from OM content), and metabolic quotient (CO_2-C h^{-1} C_{mic}^{-1}).

In order to study the decomposition rate of rice straw (first harvest), five litterbags were deposited on each plot on the soil surface in July after the first harvest. The bags with a mesh size of 1×1 mm contained 5 g dry biomass – not excluding macrofauna entirely. The litterbags were collected after 7, 15, 30, 60, and 120 days and were dried at 60 °C for 48 h and weighed. An exponential regression was applied in order to calculate the decomposition time of the rice straw.

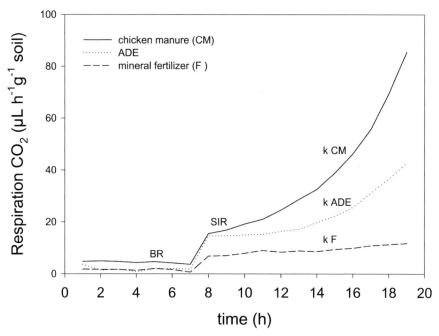

Fig. 15.2. Respiration curves of Xanthic Ferralsols amended with mineral fertilizer (*F*), soil with chicken manure (*CM*), as well as an ADE. Manure and mineral fertilizer were applied 6 months before soil sampling. Parameters basal respiration (*BR*), substrate-induced respiration (*SIR*), and the slopes (*k*) are marked. Substrate (glucose) was added after 7 h

15.2.3
Statistical Analyses

Statistical analyses were performed using Sigma Stat32 (Jandel Corporation). To evaluate the differences in the mean values among the treatments, a two-way analysis of variance (ANOVA) was performed and the Student-Newman-Keuls Method was used to detect significant differences between treatments ($p < 0.05$). The Pearson Product Moment Correlation was performed to assess the correlation between the measured parameters. The field plan was drawn with CorelDRAW (Corel Corporation) and the plots were made using SigmaPlot (SPSS Inc.).

15.3
Results and Discussion

The soil microbial population growth rate (k) after substrate addition shows a significant positive correlation to nutrient availability in the soil. The studied primary and secondary forest soils had no or negative population growth after glucose addition despite their relatively high nitrogen levels and were not included in the correlation. The highest correlation coefficients (r) and lowest p values were found for Ca ($r=0.650$, $p < 0.001$), Mg ($r=0.595$, $p < 0.001$), and K ($r=0.608$, $p < 0.001$), which indicates a pH dependency (pH in H_2O $r=0.534$ $p < 0.001$, pH in KCl $r=0.579$, $p < 0.001$). The correlation coefficient for P was 0.490 and for N 0.561 ($p < 0.001$).

The nutrient contents (K, Mg, Ca, and P) and pH values (in water and KCl) of the chicken manure treatment (C_M) are significantly different to all other treatments [excluding ADE, primary (PF), and secondary forest (SF), for which insufficient nutrient data for Mg, Ca, K, and P contents were available]. The N levels do not differ significantly between the treatments. ADE showed a significantly higher N content in comparison to all other soils ($p < 0.05$, two-way ANOVA, and Student-Newman-Keuls post hoc comparison, Table 15.2).

Excluding the chicken manure treatment, pH in water is significantly higher in treatment F (control + NPK + lime) than in the unfertilized (NPK + lime) soils (C, L, L_B, C_C, $C_C^1/_2+C_O^1/_2$, $C_C+ C_O^1/_2$). Measured in KCl the pH in treatment F and C_C+F (charcoal, NPK + lime) is significantly higher than the unlimed soils apart from treatment $C_C^1/_2+C_O^1/_2$ (the compost used for the experiment contained lime). The Mg contents are significantly increased due to liming in comparison to all unlimed treatments. Significant differences in the Ca levels are just manifested if no compost was added (Table 15.2).

The microbial population size and the microbial population growth (k) after substrate addition correlate significantly with plant biomass production (first biomass yield $r=0.754$, $p < 0.001$ and $r=0.756$, $p < 0.001$, respectively). The fourth biomass yield correlates significantly with k and microbial biomass derived from samples taken in the year 2002 after the harvest ($r=0.700$, $p < 0.001$ and $r=0.756$, $p < 0.001$, respectively). Due to their high OM and nutrient (mainly P and Ca) contents, chicken manure amended soils (C_M) showed significantly enhanced basal respiration and microbial population growth rates in comparison to all other treatments, secondary forest, and primary forest. Soil respiration curves can serve as a fast indicator of soil fertility, providing the same or similar results as plant biomass production (Fig. 15.3).

First-year respiration curves reflect clearly the level of fertilization (Fig. 15.4). BR, SIR, and microbial biomass are significantly enhanced in the plots treated with chicken manure and with mineral fertilizer (Table 15.3). The values obtained for the control coincide with those obtained by Förster and Farias (2000) in the same study area. The respiration curves after the fourth harvest are clearly diminished although to a lesser extent in the OM (charcoal)-treated plots (Table 15.4 and Fig. 15.5). The microbial growth rate after substrate addition is significantly enhanced in mineral-fertilized charcoal amended plots (C_C+F) in comparison to just mineral-fertilized plots (F). This difference supports the hypothesis that charcoal additions reduce nutrient leaching and/or microbes are able to solubilize phosphate to a greater extent. Treatment L (litter) shows a significantly increased BR in comparison to other unfertilized plots (C, C_C, $C_C^1/_2+C_O^1/_2$, and $C_C+ C_O^1/_2$). This could be

Table 15.2. Chemical characteristics of the soil samples (mean and standard deviation, SD; $n=5$). In addition, the half life of rice straw decomposition ($t_{1/2}$) is listed. Significant differences ($p < 0.05$) are marked by an *asterisk* and significantly different treatments are noted in *parentheses*

Treatment	N (g kg⁻¹)	SD	P (mg kg⁻¹)	SD	K (mg kg⁻¹)	SD	Ca (mg kg⁻¹)	SD	Mg (mg kg⁻¹)	SD	pH H₂O	SD	$t_{1/2}$ (days)	SD
C	1.82	0.14	1.87	0.67	2.66	0.61	0.97	0.21	0.77	0.15	4.50	0.15	124	24
L	1.53	0.35	1.03	0.16	2.17	0.60	1.87	0.81	0.89	0.32	4.58	0.16	118	7
L_B	1.80	0.10	2.54	0.90	2.54	0.18	3.02	2.17	1.06	0.28	4.52	0.07	106	24
F	1.51	0.20	5.15	4.66	2.44	0.68	21.0	13.0	7.69	2.95	5.04	0.25	100	17
C_M	1.85	0.24	392*[all]	420	21.3*[all]	8.95	213*[all]	121	16.4*[all]	11.2	5.58*[all]	0.64	104	14
C_C	1.73	0.27	2.91	2.34	3.66	1.50	2.91	3.24	1.29	0.83	4.53	0.19	119	21
C_O+F	1.71	0.51	8.62	7.48	2.69	1.28	26.8	15.7	6.94	3.71	4.80	0.08	80*[1,7,12]	23
C_C+F	1.65	0.10	3.41	0.80	2.77	0.76	18.4	3.75	7.20	1.82	4.71	0.34	109	23
C_C½+C_O½	1.87	0.10	5.77	4.23	3.74	0.38	11.0	2.74	2.23	0.46	4.61	0.06	113	7
C_C½+C_O½+F	1.67	0.22	7.62	2.48	3.14	0.85	23.4	18.4	7.12	3.20	4.82	0.24	99	10
C_C+C_O½	1.76	0.22	4.87	2.04	3.90	1.79	8.49	4.95	2.04	0.86	4.66	0.06	114	15
C_C+C_O½+F	2.02	0.24	4.44	3.84	3.69	1.58	21.3	15.8	4.86	4.69	4.78	0.24	96	15

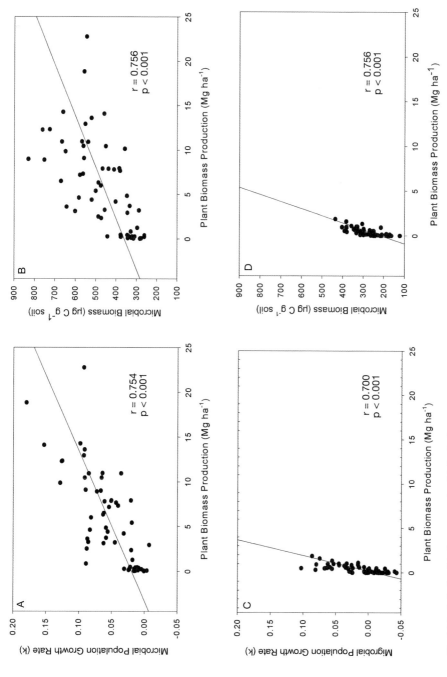

Fig. 15.3. Relationship between microbiological parameters (rate constant k and microbial biomass) and plant biomass production. **A** and **B** represent biomass production of the first harvest and **C** and **D** biomass production of the fourth harvest

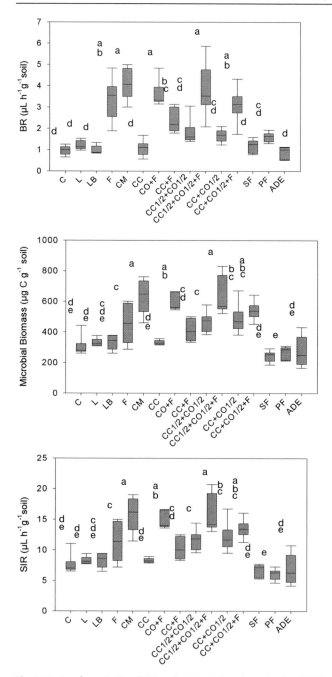

Fig. 15.4. Basal respiration (*BR*), substrate-induced respiration (*SIR*), and microbial biomass of Xanthic Ferralsols amended with different organic and inorganic fertilizers after the first year in comparison with secondary (*SF*) and primary forest (*PF*) soil as well as ADE soils. Significant differences are indicated by different *letters* (*p* <0.05)

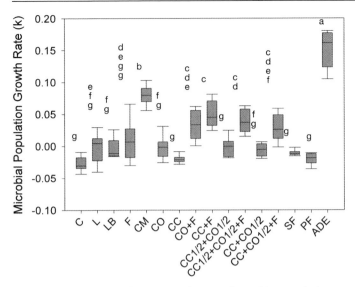

Fig. 15.5. Microbial population growth rate of Xanthic Ferralsols amended with different organic and inorganic fertilizers in the second year after the 4th harvest in comparison with secondary and primary forest soil as well as ADE soils. The reduction in *k* is highest in mineral-fertilized soil. Population growth after substrate addition is significantly higher in the charcoal-amended and mineral-fertilized treatment (*CC+F*) in comparison to the treatment mineral-fertilized alone without charcoal addition (*F*). Significant differences between the groups are marked with *letters* ($p < 0.05$)

explained either by the decomposition of the litter applied 20 months prior to measurement or by micronutrient fertilization. The effect of charcoal seems not to be a micronutrient supply, otherwise the treatment just amended with charcoal powder (C_C) could not have a significantly lower BR, and the population increase of treatment L is as low as in the other soils that were not amended with mineral fertilizer or chicken manure (C, L_B, C_O, C_C, $C_C^1/_2+C_O^1/_2$, and $C_C+ C_O^1/_2$).

The half-life of the rice litter shows a significant negative correlation with the plot's BR ($r=-0.331$, $p=0.00984$). Förster and Farias (2000) proved that litterbags buried in the soil reflect the soil's microbial activity much better than when distributed on the soil surface. The results of this litterbag study should not be overvalued, but a general trend shows that fertilized plots tend to have shorter half-life periods than unfertilized plots. Significant differences were found between the treatment amended with compost and mineral fertilizer (C_O+F, mean 80 days) and the control (C, 124 days), litter (L, 118 days), charcoal (C_C, 119 days), and charcoal+compost ($TC_C+C_O^1/_2$, 114 days).

Table 15.3. Respiration and microbial biomass of Xanthic Ferralsols with different types of organic and inorganic fertilizer amendments, forest soils and ADE in central Amazonia in the first year after the first harvest ($n=5$); significant differences are indicated by different *letters* ($p < 0.05$). *BR* Basal respiration; *SIR* substrate-induced respiration; *Biomass* microbial biomass; C_{mic}/C_{org} microbial population independent from OM content; *k* microbial population growth rate after substrate addition; *QR* activation quotient (BR/SIR); $CO_2\text{-}Ch^{-1}\,C_{mic}^{-1}$ metabolic quotient

Treatment	BR (μl h^{-1} g^{-1} soil)	SIR (μl h^{-1} g^{-1} soil)	Biomass (μg C g^{-1} soil)	C_{mic}/C_{org}	k	QR	$CO_2\text{-}Ch^{-1}\,C_{mic}^{-1}$ (Basal)
C	0.97 d	7.71 de	309.0 de	0.0147 bc	0.0101 de	0.126 d	0.00153 d
L	1.22 d	8.33 de	334.0de	0.0199 abc	0.0116 de	0.146 d	0.00177 d
L$_B$	0.10 d	8.27 cde	331.4 cde	0.0154 bc	0.0149 de	0.121 d	0.00146 d
F	3.34 ab	11.4 c	454.8 c	0.0261 ab	0.0539 bc	0.295 a	0.00358 a
C$_M$	4.10 a	15.8 a	630.9 a	0.0325 ab	0.143 a	0.265 ab	0.00321 ab
C$_C$	1.07 d	8.31 de	333.1 de	0.0155 bc	0.0408 cd	0.128 d	0.00156 d
C$_O$+F	3.66 a	14.9 ab	598.6 ab	0.0362 a	0.0916 b	0.245 abc	0.00297 abc
C$_C$+F	2.40 bc	10.3 cd	414.2 cd	0.0193 bc	0.0513 c	0.239 abc	0.00290 abc
C$_C$^{1}/$_2$+C$_O$^{1}/$_2$	1.85 cd	11.6 c	464.4 c	0.0191 bc	0.0476 cd	0.169 cd	0.00205 d
C$_C$^{1}/$_2$+C$_O$^{1}/$_2$+F	3.87 a	16.2 a	647.4 a	0.0311 ab	0.0673 bc	0.238 abc	0.00289 abc
C$_C$+C$_O$^{1}/$_2$	1.67 cd	12.2 bc	489.0 bc	0.0219 abc	0.0539 cd	0.140 d	0.00170 d
C$_C$+C$_O$^{1}/$_2$+F	3.11 ab	13.5 abc	541.0 abc	0.0187 abc	0.0674 bc	0.231 abc	0.00280 abc
SF	1.17 d	6.63 de	265.9 de	0.00512 c	-0.0109 e	0.185 bcd	0.00224 bcd
PF	1.60 cd	6.04 e	242.1 e	0.00466 c	-0.0203 e	0.266 ab	0.00323 ab
ADE,	0.87 d	6.98 de	279.9 de	0.00450 c	0.150 a	0.127 d	0.00154 d

Table 15.4. Respiration and microbial biomass of Xanthic Ferralsols with different types of organic and inorganic fertilizer amendments, forest soils, and ADE in central Amazonia in the second year after the fourth harvest ($n=5$); significant differences are indicated by different *letters* ($p <0.05$). *BR* Basal respiration; *SIR* substrate-induced respiration; *Biomass* microbial biomass; C_{mic}/C_{org} microbial population independent from OM content; k microbial population growth rate after substrate addition; *QR* activation quotient (BR/SIR); $CO_2\text{-}Ch^{-1} C_{mic}^{-1}$ metabolic quotient

Treatment	BR ($\mu l\,h^{-1}\,g^{-1}$ soil)	SIR ($\mu l\,h^{-1}\,g^{-1}$ soil)	Biomass ($\mu g\,C\,g^{-1}$ soil)	C_{mic}/C_{org}	k	QR	$CO_2\text{-}Ch^{-1} C_{mic}^{-1}$ (Basal)
C	0.62 e	4.30 de	172.4 de		−0.0269 g	0.148 bcd	0.00179 bcd
L	1.30 bc	6.27 abcde	251.2 abcde		−0.0033 efg	0.209 abc	0.00254 bcd
L_B	1.10 cde	5.65 bcde	226.5 bcde		−0.0025 efg	0.193 bcd	0.00234bcd
F	0.94 cde	6.26 abcde	251.2 abcde		0.0087 defg	0.153 bcd	0.00186 bcd
C_M	1.80 a	8.21 ab	328.9 ab	0.0129 a	0.0799 b	0.222 ab	0.00269 ab
C_O	0.84 cde	7.30 abc	292.7 abc	0.0120 ab	−0.0021 efg	0.116 d	0.00141 d
C_C	0.62 e	4.10 e	164.6 e	0.0057 ef	−0.0203 g	0.150 bcd	0.00181 bcd
C_O+F	1.34 bc	8.70 a	348.7 a	0.0108 abc	0.0330 cde	0.153 bcd	0.00186 bcd
C_C+F	1.18 cd	7.63 abc	305.9 abc	0.0095 bcd	0.0505 c	0.154 bcd	0.00187 bcd
$C_C^{1/2}$+$C_O^{1/2}$	0.74 de	5.58 cde	223.7 cde	0.0069 ef	−0.0017 g	0.141 bcd	0.00170 bcd
$C_C^{1/2}$+$C_O^{1/2}$+F	1.02 cde	8.69 a	348.4 a	0.0117 ab	0.0392 cd	0.117 d	0.00142 d
C_C+$C_O^{1/2}$	0.78 de	6.48 abcde	259.7 abcde	0.0079 de	−0.0062 efg	0.119 d	0.00144 d
C_C+$C_O^{1/2}$+F	0.97 cde	8.15 abc	326.6 abc	0.0090 cd	0.0289 cdef	0.120 d	0.00146 d
SF	1.17 cd	6.63 abcd	265.9 abcd	0.0051 f	−0.0109 g	0.185 bcd	0.00224 bcd
PF	1.60 ab	6.04 bcde	242.1 abcd	0.0047 f	−0.0203 g	0.266 a	0.00323 a
ADE	0.87 cde	6.98 abc	279.9 abc	0.0045 f	0.1500 a	0.127 cd	0.00154 cd

15.4
Characteristics of ADE and Forest Soils

Owing to a high OM content, primary forest (PF) soil has a high BR which is besides treatment L, C_M, and C_O+F significantly higher than that of the other treatments compared in the second year. Compared to treatment L (litter) and C_M (chicken manure), PF soil has a significantly increased respiratory quotient (QR) and a significantly greater carbon turnover per unit of microbial biomass (CO_2-C/C_{mic} h^{-1}) (Table 15.4).

ADE has low BR but high microbial population growth rates. The BR of ADE is significantly lower than that of soils amended with chicken manure (C_M) and PF soils. However, the population increase after substrate addition is significantly higher than those of all treatments and forest soils apart from (C_M). ADE and PF soils are both exceptional in comparison to the treatments. The plots show a positive correlation between BR and population growth rate, while it is negative in ADE and PF soils. PF soils are characterized by a relatively high BR and negative population growth and ADE by a very low BR but very high population growth after substrate addition (Fig. 15.6). These

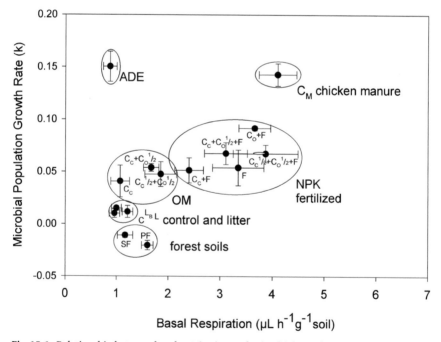

Fig. 15.6. Relationship between basal respiration and microbial population growth rate (k) after the first harvest. ADEs are exceptional with a low basal respiration (BR) and high growth rate. Primary forest (PF) soil has relatively high BR but low k. *Error bars* indicate standard error of the mean (n=5 for plot experiment; n=5 for secondary and primary forests; n=5 for ADE)

results reflect the relatively high biodegradable OM content of PF topsoil but little available nutrients in contrast to high (mean 5.8%, SD 1.4) non-biodegradable stable ADE OM with high available nutrient contents. This is a characteristic of the tropical Oxisols under PF and in contrast to the exceptional anthropogenic ADE soils containing charcoal.

Acknowledgements. The research was conducted within SHIFT ENV 45, a German–Brazilian cooperation, and financed by BMBF, Germany, and CNPq, Brazil (BMBF no. 0339641 5A, CNPq 690003/98-6). The Austrian Academy of Sciences contributed financially. We are grateful for Bruno Glasers' (University of Bayreuth) valuable advice and for the fieldworkers' help particularly Luciana Ferreira da Silva and Franzisco Aragão Simão and the laboratory technician Marcia Pereira de Almeida. Special thanks for valuable information and introduction to the IRGA-based soil respiration analyses go to Bernhard Förster (ECT Oekotoxikologie, Germany) and Florian Raub (University of Karlsruhe, Germany).

References

Anderson JPE, Domsch KH (1978) A physiological method for the quantitative measurement of microbial biomass in soils. Soil Biol Biochem 10:125–221

Beck T, Bengel A (1992) Die mikrobielle Biomasse in Böden, Teil II. Schule und Beratung (SuB) 2: III-1–III-5

Braida WJ, Pignatello JJ, Lu YF, Rovikovitch PI, Neimark AV, Xing BS (2003) Sorption hysteresis of benzene in charcoal particles. Environ Sci Technol 37(2):409–417

Fageria NK (ed) (1998) Manejo da Calagem e Adubacao do Arroz. Tecnologia para o arroz de terras altas. Embrapa Arroz e Feijao, Santo Antonio de Goias, 67–79 pp

FAO (1990) Soil map of the world. FAO, Rome

Fischer Z, Bienkowski P (1999) Some remarks about the effect of smoke from charcoal kilns on soil degradation. Environ Monit Assess 58(3):349–358

Förster B, Farias M (2000) Microbial respiration and biomass. In: Beck L (ed) Soil fauna and litter decomposition in primary and secondary forests and a mixed culture system in Amazonia. Final report of project SHIFT ENV 52 (BMBF no 0339675). Staatliches Museum für Naturkunde, Karlsruhe, pp 59–64

Glaser B, Balashov E, Haumaier L, Guggenberger G, Zech W (2000) Black carbon in density fractions of anthropogenic soils of the Brazilian Amazon region. Organic Geochem 31:669–678

Glaser B, Haumaier L, Guggenberger G, Zech W (2001) The *"terra preta"* phenomenon: a model for sustainable agriculture in the humid tropics. Naturwissenschaften 88:37–41

Glaser B, Lehmann J, Zech W (2002) Amelioration physical and chemical properties of highly weathered soils in the tropics with charcoal – a review. Biol Fertil Soils 35:219–230

Goldammer JG (1993) Historical biogeography of fire: tropical and subtropical. In: Goldammer JG (ed) Fire in the environment: the ecological atmospheric, and climatic importance of vegetation fires. Wiley, New York, pp 297–314

Goudriaan J (1995) Global carbon and carbon sequestration. In: Beran M A (ed) Carbonsequestration in the biosphere: processes and prospects. NATO ASI Series I: Global environmental change, vol 33. Springer Berlin Heidelberg New York, pp 3–18

Kimura R, Nishio M (1989) Contribution of soil microorganisms to utilization of insoluble soil phosphorus by plants in grasslands. In: Proc 3rd Grassland Ecology Conf, Czechoslovakia, pp 10–17

Kuhlbusch TAJ. Crutzen PJ (1995) Toward a global estimate of black carbon in residues of vegetation fires representing a sink of atmospheric CO_2 and a source of O_2. Global Biogeochem Cycles 9:491–501

Kuhlbusch TAJ, Andreae MO, Cachier H, Goldammer JG, Lacaux J-P, Shea R, Cruzen PJ (1996) Black carbon formation by savanna fires: Measurements and implications for the global carbon cycle. J Geophys Res 101(D19):23651–23665

Lehmann J, da Silva Jr, Pereira J, Rondon M, da Silva Cravo M, Greenwood J, Nehls T, Steiner C, Glaser B (2002) Slash and char – a feasible alternative for soil fertility management in the central Amazon? In: Proc 17th World Congr of Soil Science, Bangkok, Thailand, International Union of Soil Science, Pap 449, pp 1–12

Lehmann J, da Silva Jr, Pereira J, Steiner C, Nehls T, Zech W, Glaser B (2003) Nutrient availability and leaching in an archaeological Anthrosol and a Ferralsol of the central Amazon basin: fertilizers, manure and charcoal amendments. Plant Soil 249:343–357

Mehlich A (1984) Mehlich-3 soil test extractant: a modification of Mehlich-2 extractant. Commun Soil Sci Plant Anal 15:1409–1416

Nishio M (1996) Microbial fertilizers in Japan. FFTC-Extension Bulletins 1–12. National Institute of Agro-Environmental Sciences, Ibaraki, Japan

Pietikainen J, Kiikkila O, Fritze H (2000) Charcoal as a habitat for microbes and its effect on the microbial community of the underlying humus. Oikos 89(2):231–242

Saito M, Marumoto T (2002) Inoculation with arbuscular mycorrhizal fungi: the status quo in Japan and the future prospects. Plant Soil 244(1–2):273–279

Schmidt MWI, Noack AG (2000) Black carbon in soils and sediments: analysis, distribution, implications and current challenges. Global Biogeochem Cycles 14:777–793

Skjemstad J (2001) Charcoal and other resistant materials. In: Kirschbaum MUF, Mueller R (eds) Proc Net Ecosystem Exchange Worksh, Cooperative Research Centre for Greenhouse Accounting, Canberra, Australia, pp 116–119

Skjemstad JO, Clarke P, Taylor JA, Oades JM, McClure SG (1996) The chemistry and nature of protected carbon in soil. Aust J Soil Res 34(2):251–271

Stenström J, Stenberg B, Johanson M (1998) Kinetics of substrate-induced respiration (SIR): theory. Ambio 27(1):35–39

Trumbore SE, Chadwick OA, Amundson R (1996) Rapid exchange between soil carbon and atmospheric carbon dioxide driven by temperature change. Science 272:393–396

Uvarov AV (2000) Effects of smoke emissions from a charcoal kiln on the functioning of forest soil systems: a microcosm study. Environ Monitor Assess 60(3):337–357

Verburg PSJ, van Dam D, Marinissen JCY, Westerhof R, van Breemen N (1995) The role of decomposition in C sequestration in ecosystems. In: Beran M A (ed) Carbonsequestration in the biosphere: processes and prospects. NATO ASI Series I: Global environmental change, vol 33. Springer Berlin Heidelberg New York, pp 3–18

Wardle DA, Zackrisson O, Nilsson MC (1998) The charcoal effect in Boreal forests: mechanisms and ecological consequences. Oecologia 115:419–426

Subject Index

Printing: Mercedes-Druck, Berlin
Binding: Stein+Lehmann, Berlin